LOADING THE DICE

LOADING THE DICE

A FIVE-COUNTRY STUDY OF
VINYL CHLORIDE REGULATION

Joseph L. Badaracco, Jr.

HARVARD BUSINESS SCHOOL PRESS
BOSTON, MASSACHUSETTS

Harvard Business School Press, Boston 02163

© 1985 by the President and Fellows of Harvard College.

All rights reserved.

Printed in the United States of America.

89 88 87 86 85   5 4 3 2 1

Library of Congress Cataloging-in-Publication Data

Badaracco, Joseph.
  Loading the dice.

  Includes index.
  1. Industrial hygiene—Government policy. 2. Industry
and state. 3. Chemical workers. 4. Vinyl chloride—
Toxicology. I. Title.
  HD7269.C45B33 1985   363.1'79   85-14107
  ISBN 0-87584-162-7

CONTENTS

ACKNOWLEDGMENTS

This book is based principally on more than a hundred interviews with officials of companies, trade associations, government agencies, and labor unions in the United States, Japan, Great Britain, France, and West Germany. None of them has been quoted without permission, and the interviewees in each country reviewed my description of their organization and my account of their approach to the vinyl chloride problem.

The generous cooperation of these many public and private officials made this research possible. Among them, I am particularly grateful to Ralph Harding of the Society of the Plastics Industry in New York; John Stafford, Imperial Chemical Industries, London; John Claude Thomas, Rhône-Poulenc Industries, Paris; Haimo Emminger, Association of Plastics Producers, Frankfurt; and Dean Jiro Tokuyama and Masasuke Ide of the Nomura School of Advanced Management, and Shigeo Nakamura of the Japan PVC Association, Tokyo.

Professors Joseph L. Bower, J. Ronald Fox, George C. Lodge, and Thomas K. McCraw of the Harvard Business School have provided perceptive, stimulating, and friendly advice from the beginning of this project.

My editor, Barbara Feinberg, helped me organize and communicate my ideas much more clearly, and both Rachael Daitch and Abby Ourge assisted in the innumerable changes I have made in the manuscript.

I am also grateful to E. Raymond Corey of the Division of Research at Harvard Business School for his support and assistance, and to the Bertrand Fox Publication Fund for its financial support of the project.

LOADING THE DICE

INTRODUCTION

This study describes how major chemical companies and government agencies in the United States, Japan, Great Britain, France, and West Germany resolved the same dramatic, complex problem: reducing the exposure of workers to vinyl chloride (VC), an invisible, carcinogenic gas used to make plastic products that are found in virtually every home, office, and factory in the developed world. Like the tragedy in Bhopal, India, and other chemical catastrophes, the VC episode—which has caused at least 108 deaths—raised the urgent, practical issues of limiting human exposure to the toxic substances routinely used in vast quantities in modern industrial operations.

But the VC episode took a peculiar form. The five countries examined in this study all learned of the VC-cancer link at the same time, in January 1974. All took action immediately, starting out with essentially the same fragmentary information about the technology and cost of VC control. In all the countries, major labor groups, such as the AFL-CIO in the United States and the Chemical Workers Union in West Germany, confronted powerful companies, including ICI in Great Britain, Mitsui Chemicals in Japan, Rhône-Poulenc in France, and Dow, Shell, and Union Carbide in the United States. In every case a government agency responsible for workplace health and safety became deeply involved with the problem. And, in the end, all five countries reduced VC exposures to the same very low levels.

What differed among the countries were the actions taken by business and government to resolve the problem. Hence, the VC episode proved to be very nearly a laboratory experiment: the basic issue—limiting VC exposure—held constant across all the countries, thereby highlighting differences in business and government relations. The VC problem casts light on three important questions:

- How do business-government relations differ from country to country?
- Why do they differ?
- Do the differences matter?

Responses to these questions lie at the intersection of a wide range of important problems, both practical and intellectual. American business-government relations are often criticized —by Americans and others. Proposals to improve U.S. productivity, international competitiveness, and environmental, workplace, and consumer safety regularly recommend greater business-government cooperation. Many business executives and government officials advocate collaboration because of their own frustrating experiences with "adversarial" relations and because they believe that in other countries—such as Japan, West Germany, and some of the newly industrializing countries—business and government have successfully joined forces. For political scientists, political economists, and analysts of regulation and its reform, a comparative approach to a problem like the VC episode may reveal questions about national decision making, pluralism and corporatism, comparative politics and regulation, and the role of technology and technocrats in complex decision making.

This study concentrates on how institutional arrangements can encourage cooperation between business and government on complex, controversial problems like the VC problem. The reason is that a clear pattern emerged from the VC "experiment." In Japan and the three European countries, public and private officials worked together for several years to resolve the VC problem. Their cooperation took essentially the same form, despite strong differences in national history and culture, and rested upon fundamentally similar institutional arrangements linking business and government. In contrast, the U.S. approach to the problem was unambiguously adversarial: business and government did not work together but instead expended considerable effort seeking to restrain, impede, and discredit each other. Underlying the conflict were institutional arrangements that differed dramatically from those in Japan and Europe.

Institutional arrangements are not the sole explanation of cooperation and conflict. An omniscient social scientist would, no doubt, explain business-government relations in a long regression equation with variables for history, political culture, ideology, political economy, and other factors. This study suggests a more limited conclusion: namely, that institutional arrangements—in Max Weber's phrase—load the dice in favor of cooperation or conflict. Moreover, this systematic comparative study of different approaches to an important, complex, and controversial problem may contribute to more pragmatic and successful changes in the way companies and government agencies deal with each other in the United States.

Chapter 1
THE VINYL CHLORIDE PROBLEM

The control of vinyl chloride was one of the major international regulatory episodes of the 1970s. Virtually overnight, a glamorous, prosperous, and fast-growing industry was imperiled by evidence that linked its principal raw material to incurable liver cancer. Workers who handled VC were at risk; grave questions were widely publicized about the health of consumers of polyvinyl chloride (PVC) products and the general public. Throughout the world by the late 1970s, vinyl chloride routinely appeared on lists of major health hazards. Its companions included asbestos, coke oven emissions, DDT, PCBs, cyclamates, red dyes #2 and #4, Tris, Kepone, nuclear radiation, carbon black, chloroform, diethylstilbestrol, arsenic and yellow dye #1. By 1980, the major VC-producing countries had regulated VC as an air pollutant and as an occupational health hazard. Its use as an aerosol propellant had generally been banned.

In the United States, measures to resolve the VC problem were devised under government-imposed "emergency" procedures, amid allegations of crisis and cover-up, and under the glare of intense media scrutiny. At the first meeting among representatives of industry, labor, and government, who had gathered to discuss the VC issue, harsh controversy erupted, and it continued even after industry had complied fully with the final regulations for VC. At one point, senior union officials accused the VC and PVC companies of blackmail, a government official described VC as the "occupational disease of the century," and *Time* magazine featured an article entitled "The Plastic Peril." Powerful players were arrayed on all sides. The Occupational Safety and Health Administration (OSHA), backed by the National Institute for Occupational Safety and Health (NIOSH), took the lead government role. B.F. Goodrich, Union Carbide, Shell, Conoco, and Dow Chemical, along with a major industry association, formed the industry side. The AFL-CIO, the United Rubber Workers, the United Steel Workers, and the Oil, Chemical and Atomic Workers represented most of the 6,500 workers in VC and PVC plants and many of the 700,000 workers who fabricated products from PVC resins.

In contrast to this adversarial approach, business, government, and labor unions in Great Britain, France, West Germany, and Japan—all facing the same VC problem—cooperated with each other in one fashion or another to solve the problem. To be sure, each country had its own approach, but what is striking are the similarities in the ways these four countries handled the issues raised by VC and the differences that separate their approach from that of the United States.

The VC episode represents a large class of important problems that companies and government agencies will confront again and again in all industrialized countries. In most respects, the VC situation was typical of a broad range of problems such as those involving industrial pollutants, food additives, pharmaceutical products, occupational health and safety, and nuclear reactor operations. These problems originate in industries sharing three fundamental and ostensibly unremarkable characteristics: the industries use high technology to manufacture virtually ubiquitous products; misuse of the products or mismanagement of the industry can cause severe and widespread social harm; and enormous practical complexities impede the resolution of these problems.

Ubiquitous High Technology Products

Almost everyone reading this page can reach out and touch an object made of polyvinyl chloride. PVC is used in thousands of consumer and producer products. These include vinyl flooring, phonographic records, upholstery, wire and cable insulation, food wrapping, auto bodies, baby pacifiers, telephone equipment, credit cards, conveyor belts, plastic bottles, coating for pollution control equipment, shoes, and medical devices. The protean diversity of its uses results from PVC's low cost, its ability to resist fire, rust, and corrosion, and the ease with which it combines with other materials to meet a wide range of industrial and consumer needs. It is even possible to buy a vinyl-based garden hose that gives off a cherry scent. Further, PVC is low in cost, resists fire, rust, and corrosion, and is easily combined with other materials to make an even wider range of consumer goods. From the early 1950s onward, world PVC production grew at rates approaching 15 percent a year, and during the 1970s an average citizen in an industrialized country consumed more than 20 pounds of PVC products annually.

Behind this explosive growth was sophisticated technical expertise: each step in the production of VC, PVC, and PVC plastic products required solving complex problems in chemistry, physics, and engineering. VC is composed of simple molecules called monomers, and for this reason, VC is also called VCM or vinyl chloride monomer. VC is made into PVC in a process called polymerization, which makes large complicated molecules out of the simpler molecules of a monomer such as VC. Polymers are the chemical building blocks of the modern plastics industry.

Four different technologies are used to make PVC from VC. In each of them, the VC is mixed with other substances, called catalysts, in a large vat or reactor. Then the mixture is heated and agitated. The result is a crude polymer that may be further treated with plasticizers, flame retardants, colorants, stabilizers, and other materials to produce a final blend, suited to a particular customer's needs. PVC pipe and baby pacifiers require different types of PVC, and several hundred polymer blends are available commercially.

The last step in making an industrial or consumer product from PVC is fabrication. A polymer blend is heated, additional materials may be added, it is subjected to high pressure, and then formed into final products by extrusion, spraying, injection-molding, calendering, casting and other techniques.

Like PVC plastics, any chemical product that reaches market in the late 1980s is almost certainly the result of a decade or more of collaboration among chemists, physicists, statisticians, biologists, and engineers. In fact, it is misleading to simply list these disciplines. Each is a kingdom unto itself, within which subdisciplines branch into increasingly refined specializations, each with its own elaborate body of knowledge and technique. Increasingly common among these specialists are experts on health and safety. Epidemiologists, industrial hygienists, toxicologists, pathologists, and others routinely participate in company and government decision making and attempt to assess the health effects of industrial products. Other experts use techniques like cost/benefit analysis to estimate broader social and economic effects.

This ascendancy of specialists who assess the social impact of industrial production results from the second fundamental feature of modern industrial activity. Misuse of the products or mismanagement in the industry may cause widespread and severe social harm. The economic side of this social damage includes slow growth, layoffs, regional decline, and so forth. The health side ranges from minor injuries to cancer, blindness, sterility, and death. On all sides, the stakes are high—as was clearly the case with VC.

High Social Stakes

There were many points in the VC-PVC production process at which workers could be exposed to VC. At room temperature VC is a gas but is stored and shipped under pressure as a liquid. Since it evaporates as soon as it is released from the pressurized vessel, it is very hard to control: small leaks can go unnoticed while gas disperses. As a result, workers in VC plants could be exposed in at least four ways: through these "fugitive" leaks, in loading VC into railroad tank cars that carry it to PVC plants, in quality control sampling and analysis, and during the maintenance and repair of vessels used to transport and store VC.

In polymerization plants, exposure could occur during the manual cleaning of the reactors in which the polymerization occurs. Cleaning is necessary because patches and chunks of PVC accumulate on the inside walls of the reactor and interfere with subsequent polymerization. An anomaly of PVC production was the reliance, in a technologically sophisticated industry, on a reactor cleaning technique—the manual use of a hammer and chisel—invented during the Iron Age. Reactor cleaners were exposed to VC because some of the gas remained in the PVC residue and was released by the chiseling. Exposure could also occur during the unloading of VC tank cars via fugitive leaks from the pipes, valves, gaskets, and compressors through which VC moves at high pressure; during opening of the production system for inspection, repair, or maintenance; and during disposal of bad batches of PVC.

During the 1960s, VC became linked to two chronic health problems. One, called acroosteolysis or AOL, appeared as a broadening and flattening of the fingertips and resulted from bone deterioration. The other, Raynaud's syndrome, was characterized by an abnormal sensitivity to cold and a feeling of "pins and needles" in the hands.

Discovering the link between VC and AOL led to the first of a series of studies, starting in the late 1960s, that ultimately linked VC to cancer. An Italian professor, P.L. Viola, exposed 25 rats to 30,000 ppm (parts per million) of VC for one year in an attempt to induce AOL. His attempt failed, but he did find cancers of the skin, lung, and bones. In May 1970, he reported his findings to the 10th International Cancer Congress in Houston with the conclusion that "no implications for human pathology can be extrapolated from the experimental model reported in this paper."[1]

Viola's findings prompted experiments on a possible VC-cancer link by Professor Cesare Maltoni of the Institute for Oncology in Bologna, Italy. By January 1973, Maltoni had preliminary findings of tumors of the ear, kidney, and liver at levels of 500 and 250 ppm, but not at 50 ppm or in the control animals.[2]

The first of VC-related cancers in humans were noticed near the end of 1973. On December 18, a physician at B.F. Goodrich's Louisville, Kentucky, PVC plant told Dr. N.M. Johnson, Goodrich's director of environmental health, that one Goodrich employee had recently died from angiosarcoma of the liver and that another died of the same rare disease a year earlier. Dr. Johnson had recently attended a Manufacturing Chemists Association meeting and had heard an interim report on Maltoni's study. Suspecting a connection, he started a search of employee health records and work histories.

Within a week, a third Goodrich PVC worker had died. An autopsy found angiosarcoma of the liver to be the cause of death. After taking further steps to confirm the three cases, B.F. Goodrich, on 22 January 1974, informed the National Institute for Occupational Safety and Health, the Kentucky Department of Labor, workers in its VC and PVC plants, and the national news media that VC—a cornerstone of the world plastics industry—was apparently linked to cancer in humans.

In 1974 worldwide investment in plant and equipment for VC-PVC industry totaled $8 billion. In each of the major VC-producing countries, 5,000–10,000 workers were directly involved in producing VC or PVC, and several hundred thousand jobs depended on PVC fabrication. Thus, the social stakes in the VC case went beyond the health of the workers and consumers and extended into a vast number of jobs, investments, and products—all imperiled by the possibility that VC production would be substantially curtailed or halted altogether.

Attempts to assess the social stakes were handicapped by profound ignorance about the dose-response curve for VC. Such a curve would have shown how various levels of VC exposure (dose) affected the incidence of cancer (response). Without this information, it was impossible to predict how many lives would be saved by lower VC exposure levels. On one side, industry officials argued that the VC-related deaths discovered in the 1970s were due to extremely high exposures, primarily among reactor cleaners, two or more decades earlier. Critics of this view responded that there was no empirical evidence for the existence of a "no-effect" threshold, that is, a level of exposure below which no harm occurred. Therefore, they argued, exposure to even a few ppm of a carcinogen should be prohibited. And even if the number of lives saved or statistical angiosarcoma deaths averted were known, there was and is no generally accepted way to value lives spared and pain and anxiety reduced in a unit (such as dollars) that permits direct comparison with the cost incurred.

Even under the best of circumstances these dose-response curves are very difficult to construct. Curves based on animal data raise the problem of mouse-man extrapolation. Is a mouse more sensitive or less sensitive to VC than a man? Is a two-year-old mouse, dead from liver cancer, comparable to a 60-year-old man or a 120-year-old man? Are rodent noses better or worse than human noses at filtering air-borne hazards? Thus, even if a laboratory experiment produced a clear dose-response curve, serious questions would remain about the corresponding curve for humans.

Sometimes data are available from what are, in effect, human experiments: that is, from the exposure of humans to hazards, through negligence or ignorance. Epidemiologists are medical specialists who study exposed human populations, compare them to control groups, and then draw statistical inferences about the nature and incidence of hazards. Unfortunately, many factors impede epidemiological studies: poor or nonexistent records, ambiguous death certificates ("heart failure complicated by liver cancer"), and the long latency period—roughly twenty years for VC—between exposure to a hazard and the effect under study. Occupational disease can also mimic disease from other causes. Had VC caused lung cancer, instead of a rare liver cancer, its hazards might have remained undetected for years. Finally, industrial production is not organized in neat gradations of exposure level—0, 10, 50 ppm—which, along with excellent records, would permit a dose-response curve for humans to be constructed.

Engineers, economists, accountants, and policy analysts added further uncertainties about both the costs and benefits of controlling VC exposure. Before 1974, VC exposure levels had been in the hundreds of ppm. As a result, in early 1974, there was no hard information on the costs of cutting exposure to minuscule levels in the very low double or even single digits. (One part per million is the equivalent of one second in nearly 280 hours, or one inch out of 16 miles.)

Estimating the direct cost to industry of cutting VC exposure was an elaborate exercise in speculative cost accounting. It required estimates of the cost of computerized monitoring devices and alarms, operating procedures, and technology—such as mechanized reactor cleaners—that did not exist. Indirect costs of various VC exposure levels were even harder to assess: how would as yet unknown changes in the prices of hundreds of PVC resins affect jobs, consumer prices, and market shares in the vast array of VC-dependent industries? The full social stakes of the VC problem, while undoubtedly high, proved to be quite elusive.

Practical Complexities

Reducing VC levels in plants required a painstaking, job-by-job and area-by-area analysis of possible points of exposure; and, technical challenges did not end once all the leaks in the production system were "plugged." Companies had to develop technology to monitor minute VC concentrations in workplace air. These systems had to sample the air at dozens of points in the plants, immediately cycle the samples through analytical devices, and set off alarms if dangerous exposure levels appeared. Ultimately, two established analytic methods were used, spectrophotometry and gas chromatography, but they had to be adapted for special use in VC plants.

Outside the companies other parties were grappling with other complexities. Public agencies, for example, had to determine just how exposure levels were to be calculated. The European Economic Community, for example, developed the equation, $\sigma^2\,(\tau,\ T)\ =\ 2.5\ \times\ 10^{-2}\,(\tau/t)$, which described the distribution over time of airborne VC concentrations in a large sample of European PVC plants. Researchers faced other difficulties in establishing protocols for VC experiments, since they had to test for differences in reactions among rats, mice, and other lab animals, differences among various strains of rats and mice, and different routes of administration of dose (inhalation, ingestion, intraperitoneal injection versus subcutaneous injection).

All this was closely paralleled by extreme political complexity, for the range of interest at stake was boundless: jobs, trade balances, energy, safety, and health—for workers and the public, technological innovation, inflation, and so forth. Questions of women's rights even emerged since some studies linked VC to fetal damage: did this mean that certain jobs in the chemical industry should have been closed to women of child-bearing age? Political complexity was a transnational phenomenon as well. The EEC Council of Ministers, the Conseil Europeen des Federations de l'Industrie Chemique (a European industry body representing national chemical industry associations), and the International Chemical Workers Union all became involved in the development of EEC policy, with which the national policies of member countries had to be harmonized. The political calculus became even more tangled because the major competing blocs—such as industry and labor—were often divided internally.

The technical and political complexities led, in turn, to high managerial complexity. Companies and government agencies in all countries had to collect, analyze, and disseminate data on the full range of complicated and uncertain matters relating to VC. Then, under the political spotlight, the agencies had to develop a set of workable, and politically realistic regulations, and the VC and PVC companies had to design, procure, install, and debug their VC control and monitoring systems.

The difficulty of performing these intrinsically complex and uncertain tasks was heightened by the powerful social passions that modern industrial problems can elicit. Ubiquitous products and high social stakes bring powerful emotions into play on a wide scale. The political manifestations of these emotions, and their exploitation, make these problems even harder for business and government to resolve. The VC case provides a clear illustration: the web of fears that links chemicals and cancer.

In its popular form, the link is a syllogism: chemicals equal cancer, cancer equals death, hence chemicals equal death. This grim and controversial logic touches deep fears, manifested everywhere. In the United States, Europe, and Japan, media coverage of cancer stories is intense—nearly every issue of the American daily papers carries an article on the causes, prevention, incidence, cures, or victims of cancer. During the 1970s, the budget of the National Cancer Institute grew nearly twice as fast as that of the National Institutes of Health. Over the same period, many environmentalists shifted their attention from nature laid waste to human disease. Nearly twenty years after the publication of *The Silent Spring* and *The Sea Around Us* by Rachel Carson, the best-selling environmentalist critique was probably the *Politics of Cancer* by Samuel Epstein, M.D.

Since VC was linked to an incurable cancer, its regulation was pervaded by fears and uncertainty. There was concern at the start about the hundreds of thousands of workers who fabricated PVC plastics and might be exposed to residual VC in PVC resins. Millions of people lived in the neighborhoods around VC and PVC plants, and at least one published study raised fears of birth defects among their children.[3] Consumers were also at risk because VC aerosol propellants, used in many home products, were found to cause temporary exposures as high as 400 ppm.

In the United States, Europe, and Japan, efforts to reduce

VC exposure took place while studies of work histories and
medical records steadily raised the toll of suspected VC-related
deaths. In France, a major left-wing trade union published
articles charging that workers were used as "common guinea
pigs" and suggested that wine in PVC plastic bottles was
dangerously contaminated. A Japanese consumer group issued
a report called "Fear of VC." In the United States, *The New
Republic* published an article called "Cancer in a Spray Can,"
and CBS broadcast an hour-long, prime-time special on VC
called "The Case of the Plastic Peril."

Industry and government can usually do little to calm these
fears. This is because modern industry operates on such a vast
scale that errors and breakdowns cannot be completely elimi-
nated, and some errors and breakdowns lead to tragic and
highly publicized disasters. These, in turn, erode the credibility of the industries and of the government agencies that regulate them. For example, every year, chemical companies
safely perform literally millions of intricate operations involving dangerous materials. Every year, government agencies
conduct thousands of thorough, uncontroversial inspections of
chemical plants. Yet the public view is strongly shaped by
extraordinary, highly publicized failures. All around the
world, regardless of industry efforts, there occur a steady series
of toxic releases and fatal explosions. In the spring of 1980,
while the Love Canal evacuation was taking place in the
United States, nearly a thousand residents of a Dutch housing
development were evacuated because toluene and xylene,
waste products from paint manufacture, had contaminated
nearby ground water. Almost every year, a train derailment
causes leakage of VC from rail tank cars. Some incidents
involved explosions or fires; others required evacuation of
thousands. The disaster in Bhopal, India, in December 1984
killed more than 2,000 people.

Such episodes raise the question of whether profits or bureaucratic routine have supplanted the concern for health and safety that industry and government avow. Private greed and public negligence no doubt account for many mishaps and tragedies. But there remains a significant number of serious incidents that occur simply because, out of millions of complex operations, some inevitably will fail.

When industries and government agencies with tarnished histories approach a problem like VC, scientific complexities and uncertainties further undermine their credibility. Frequently, industry must estimate, on the basis of very limited experience, the cost of reducing exposure to a hazardous substance. Uncertainty may tempt companies to bias their estimates on the high side. Even if industry resists the temptation, its critics can nevertheless assert that industry has exaggerated its estimates, and until the costs are finally incurred, there is no way for industry to prove its case. When the costs are finally known, they may be higher than the government's estimate— in which case government is likely to be charged with regulating beyond an appropriate balance of cost and benefit; or costs may be lower than the industry estimates, and industry will be accused of having "cried wolf," which is precisely what happened when VC was regulated in the United States.

Medical uncertainties can also fuel fears and impair credibility. Studies take months or years to complete. Interim results may suggest a possible hazard, but confirmation may require completion of the study. When industry sponsors the studies (and the chemical industry in the United States and Europe sponsored virtually all of the studies that revealed the VC hazard), and when the studies ultimately confirm a serious hazard, industry may be accused of a cover-up. If industry informs government of interim results that are suggestive but not conclusive, the agencies, if they do not act immediately, may later be charged with foot-dragging, proindustry leanings, or indifference to the worker or public health. These suspicions and accusations all gain plausibility from the exceptional but well-publicized cases of clear government ineptitude and industry neglect.

Again and again, industries and government agencies around the world must resolve problems that threaten their credibility, that arouse powerful social passions, that present formidable technical, political, and managerial complexity, and whose basic dimensions are highly uncertain. All these characteristics pervaded the VC problem because VC—like so many other modern industrial products—is a virtually ubiquitous, multiscience product with broad social impact.

Chapter 2
HIERARCHIES AND NETWORKS

The B.F. Goodrich announcement connecting cancer and VC confronted all five countries examined in this book with the same complex, controversial issues. All five undertook urgent efforts to control VC at the same time, and all had basically the same information about the VC hazard and the technology and cost of controlling it. Moreover, the same parties—plastics companies, government agencies, labor unions, the media, and independent scientists—were active in every case. Yet, a clear pattern cuts across the five countries. In Japan and Europe, industry and government cooperated to resolve the VC problem. In the United States they did not. Moreover, cooperation in Europe and Japan took the same basic form—despite marked differences in culture, history, and government.

The institutional arrangements linking business and government provide the strongest explanation of this pattern. In general, in Japan and in the three European countries—Great Britain, France, West Germany—the branches and agencies of government, as well as the VC and PVC companies, were organized into stronger hierarchies than their U.S. counterparts, and these strong hierarchies were linked by networks of private, informal, personal contacts among members of the public and private hierarchies. The broad framework of networks and hierarchies sketched below introduces the administrative arrangements that the five case studies describe in detail, and later in this book they provide a strong explanation of the patterns of conflict and cooperation that distinguish the five countries examined in this study.

Hierarchies

In fact, the monoliths "business" and "government" do not exist and therefore cannot have relations with each other. Moreover, neither business nor government is simply an aggregation of undifferentiated units. Instead, the two sides of a so-called business-government relationship consist of individual companies and individual branches, departments, and subsections of government. These units are organized into hierarchies that may range from strong to weak. Companies, for example, may go their separate ways in a highly competitive industry and organize themselves infrequently, tentatively, and only briefly to confront common threats. Or, under the leadership of a powerful trade association or several dominant firms, companies in an industry may routinely coordinate their economic and political efforts. The government side of a relationship is a strong hierarchy if a single agency or branch can act with limited interference from other agencies and branches, if it controls the resources its mission requires, if it can pursue consistent operating practices and maintain a stable organization, and if it has capable, respected, experienced leadership.

In all four cases of business-government cooperation on VC, government was a strong hierarchy. In every case, the Ministry of Labor was clearly first among equals: opportunities for judicial review of its decisions were extremely limited, and the ministries could anticipate support from the legislature since all four countries had variations of parliamentary government. Although other branches of the government played indirect roles, a single branch of government exercised the predominant influence on the VC decisions, and its influence, moreover, was strengthened by a long national tradition of executive branch power.

In the United States, the government side in the VC case differed dramatically from the strong Japanese and European hierarchies. During the 1970s, OSHA and the other new regulatory agencies were subject to intense congressional oversight and to almost continuous judicial review. Of course, cooperation in the other countries limited judicial review to the extent it succeeded in settling controversial issues and satisfying aggrieved parties. But in Japan and in the three European countries, there were not only fewer incentives to seek judicial review, there were also fewer opportunities to do so, and this *limited* judicial review greatly strengthened the government hierarchy in the cooperative cases. The head of the West German association of plastics producers exaggerated only slightly when describing his response to a government decision that he or his association found unacceptable: "I suppose I would have to vote differently at the next election." In the United States, an industry group and each of its member companies had wide opportunity to seek judicial review of unfavorable executive branch decisions, an opportunity they shared with labor unions, public interest groups, state and local governments, and many other parties.

Also, in Japan and in the three European countries, career civil services reflected and reinforced the power of executive agencies. Nearly 200 years later, de Tocqueville's observation remains accurate:

> In the United States, as soon as a man has acquired some education and pecuniary resources, either he endeavors to get rich by commerce or industry, or he buys land in the uncleared country and turns pioneer. . . . among most European nations, when a man begins to feel his strength and to extend his desires, the first thing that occurs to him is to get some public employment.[4]

Government careers in Great Britain, France, Japan, and West Germany have attracted an elite: the top graduates of Oxford, Cambridge, the University of Tokyo, the Ecole Polytechnique, and so forth. Their civil services have long, prestigious histories of authority and influence at the center of national affairs. The French administrative lineage runs to Napoleon or even to Colbert, the German to Prussia before Bismarck. In these countries, a civil service career is the model professional life.

Furthermore, executive branch leadership tends to be more stable and continuous in these countries, in contrast to the United States where roughly one-sixth of the federal officers and employees are not subject to the competitive requirements of civil service laws and are appointed by the president and his appointees. In the words of one observer, "Americans, even today, regard a change in government as the occasion for a veritable upheaval of the employees in the upper levels of the administration." In Great Britain, however, only the very apex of the government hierarchy changes when the government changes. This difference was apparent during the first five years of OSHA's existence, for example. There were four secretaries of labor and three assistant secretaries for OSHA. On two occasions the OSHA directorship was vacant for several months. Both the assistant secretary of labor for OSHA and OSHA's director of standards had held their positions for roughly a year at the time of the VC problem. In contrast, the Japanese and European agencies responsible for VC were headed by career civil servants. The two senior British health and safety officials had been members of the factory inspectorate for over thirty years when the VC problem arose.

These differences in the stability of leadership partly reflected the ages of the different agencies. At the start of the VC episode OSHA had existed for only five years; before that, federal activity in workplace health and safety had been confined to government employees, businesses with federal contracts, and just a few industries like mining, construction, and maritime. By contrast, the British Factory Inspectorate was nearly 150 years old when it became involved with the VC problem. In France and West Germany, the ministries of labor had been responsible for workplace health and safety since the beginning of the century. The Japanese Ministry of Labor was established in 1947 by the U.S. Occupation as part of its program to promote trade unionism in Japan, but workplace health and safety issues had been a federal responsibility before the war and had been handled by a branch of the Ministry of Health and Welfare.

The continuity of organization and operating practices on workplace health and safety also indicated the strength of executive agency hierarchies in the cooperative cases. For decades in the European countries and in Japan, inspectors had acted both as advisers and law enforcement officials. In contrast, OSHA's establishment in 1970 signaled abrupt changes in organization and operating policies. First, responsibility shifted from the states to a new federal organization, and during its first five years, the federal OSHA office had been reorganized three times. Second, state policies of cooperation with industry were replaced by new OSHA policies. In fact, objections to the generally cooperative approach taken by state agencies were one of the principal reasons for OSHA's creation. One study described the state programs in these terms:

> In general, prior to the passage of OSHA Act, state enforcement of health and safety standards (1) involved too few and insufficiently trained personnel, (2) was encumbered by a slow-moving state bureaucracy, (3) was not infrequently corrupted by abusive discretionary authority, which accompanies the vague and confused state standards, (4) did not include first-instance citations, but depended on warnings, and (5) was heavily safety-biased to the detriment of acute health problems and the virtual exclusion of chronic occupational disease.[5]

Through the 1970s, OSHA's basic operating policies shifted as administrations and Congresses changed and as court decisions were handed down. These were years of intense controversy and litigation over the rights and responsibilities of inspectors, the need for search warrants before entering company property, and over the role of cost/benefit analysis in OSHA's deliberations. Another weakness in the American administrative hierarchy responsible for workplace health and safety was that NIOSH—OSHA's scientific arm—was attached to another government body, the Center for Disease Control, headquartered in Atlanta and part of the Department of Health, Education and Welfare (now the Department of Health and Human Services). In contrast, the government agencies responsible for workplace health and safety in Europe and in Japan had, as part of their administrative hierarchy, their own captive scientific branches.

Thus, the experience and caliber of civil servants, a long tradition of centralized executive branch authority, relatively limited judicial review, and stable organizational design and practice all reflected and reinforced the positions of primacy held by the Ministries of Labor in the four cases of cooperation. In the United States, the government hierarchy involved in the VC problem was weak. Moreover, the same pattern of weak and strong hierarchies appeared on the industry side of the VC issue.

Table 2-1 introduces the industry hierarchies in the five countries.

Table 2-1　Industry Hierarchies in the United States, Great Britain, France, West Germany, and Japan

	United States	Great Britain	France	West Germany	Japan
Number of VC producers	9	2	3	6	18
Number of PVC producers	20	4	4	7	18
Percentage of national VC production capacity of the three largest VC producers	58%	100%	100%	79%	33%
Percentage of national PVC production capacity of the three largest PVC producers	29%	95%	100%	69%	29%

In Great Britain, France, and West Germany, there were relatively few VC and PVC producers, and the three largest producers in each country were responsible for a much greater fraction of national VC and PVC output than their counterparts in the United States and Japan. In fact, these data actually underestimate the degree to which the VC and PVC industries in Europe were dominated by large producers because at the time of the VC problem in France and West Germany, some of the largest producers owned major shares of other producers. Rhône-Poulenc, the largest French VC producer, owned half the shares of Société DAUFAC, the second largest producer. In West Germany, Hoechst, the largest VC producer and the third largest PVC producer, owned half of Wacker Chemie, another of the largest VC and PVC producers. In Europe, moreover, the large producers of VC were nearly always large PVC producers. In the United States, on the other hand, the two largest VC producers, Dow Chemical and Shell Chemical, did not produce PVC. In 1974, of the nine American VC companies, only four produced PVC, and only four American PVC producers produced VC.

Another factor that added to the power of the large European producers is that the parent companies of some of the largest VC and PVC producers were among the very largest companies in their countries. In 1974 British Petroleum and ICI were the two largest companies in Britain (as measured by sales). Hoechst and BASF were the third and fourth largest companies in West Germany, and the largest West German company, VEBA, owned a quarter of Huls, the country's largest VC and PVC producer. Rhône-Poulenc was the eleventh largest company in France.

These huge chemical companies exercised their power in part through industry associations. These were financed through members' contributions, which were generally proportionate to a member company's sales; hence, the large companies tended to dominate. An association dominated by a few large companies can strike bargains and coordinate efforts among its members more easily than one with a large and heterogeneous membership. Furthermore, the large companies had extensive contacts with government and labor union officials, independent of the association, which increased their leverage over smaller association members. Finally, the very large companies had operating experience, scientific and engineering expertise, research facilities, and staffs that provided them with better data and greater authority in intraindustry deliberations.

There are exceptions to this harmonious picture, since battles for competitive position can disrupt the coordination of industry hierarchies by dominant producers. But, in general, the European trade associations did reflect and reinforce a strong hierarchical structure among member companies, and they were clearly distinguishable from the temporary, weak alliances in the U.S. VC industry. If anything, American companies at the time of the VC issue were moving toward an even more fragmented approach to government relations. During the 1970s there was a substantial increase in the number of political offices established in Washington, D.C., by individual companies—a clear sign of their desire to pursue political strategies that were independent of their industry and its trade association. Moreover, the deregulation of major American industries and the ensuing competition also weakened political-economic alliances in many industries.

Turning from the European VC-PVC industry to the Japanese, Table 2-1 indicates parallels with the U.S. industry. Both the United States and Japan had many producers—in fact, there were more VC and PVC producers in Japan than in any other country—and the largest Japanese companies manufactured an even smaller fraction of total industry output than their American counterparts. Furthermore, like the American producers, the Japanese did not display the pattern of interlocking ownership that appeared in France and West Germany.

Yet these similarities belied fundamental differences between the American and Japanese VC-PVC industries. In the first place, many of the Japanese producers were linked as members of the loose economic groupings called *keiretsu*, the contemporary progeny of the powerful *zaibatsu* that dominated Japanese industrial organization until after World War II. The zaibatsu—literally "money-cliques"—were large conglomerates whose member companies held powerful positions in manufacturing, trade, transport, banking, and insurance.

Activities of each zaibatsu were coordinated both through central holding companies and family ownership. The U.S. Occupation Authority dissolved the zaibatsu after World War II,
but the original groupings soon reemerged, though the links
among the companies were looser and less formal. The prewar
pattern of control was nevertheless reflected thirty years later in
the stock ownership of several major VC and PVC producers.
For example, Mitsui, along with Mitsubishi, Sumitomo, and
Yasuda, were the four major zaibatsu before the war. Thirty
years later, at the time of the VC problem, four of the major
VC and PVC producers were associated with the Mitsubishi
group and four others were associated with Mitsui. In particular, three of the seven largest shareholders in the Toagosei
Chemical Industry were Mitsui and Company, Mitsui Life
Insurance, and Mitsui Capital Bank. Together, they held
about ten percent of Toagosei's shares.

In the late 1970s, the strength of keiretsu relations was again
manifested in the Japanese VC-PVC industry, when the industry faced overcapacity and heavy losses. The government
permitted companies to establish a number of "depression cartels"—under the auspices of the Japan PVC association—to
coordinate investments, marketing, and capacity reductions.
The companies organized themselves into four groups whose
members paralleled the original zaibatsu groupings. In Europe
as well, by the late 1970s, such "depression cartels" had also
been proposed for the petrochemical industry, which had been
staggered by competition from new plants in the Middle East,
a demand plateau, and extensive overcapacity.

The practice of cartels, combines, and other varieties of
intraindustry economic collaboration was hardly a novelty for
Japanese or European industries. It is another factor that reinforced the industry hierarchies in Japan and Europe and distinguished them from the United States. German cartels for
controlling prices and output first appeared in the seventeenth
century. During Germany's industrial revolution, cartels pro-

liferated in coal, chemicals, iron, and other industries, and these agreements were among the strongest in Europe, in part because cartel agreements were enforceable in court. Perhaps the most notorious of all German cartels was IG Farben. It was initially a trust for the dye industry and dominated the German chemical industry for nearly a third of a century, until after World War II. Its modern offspring are the current giants of the West German chemical industry—Badishe, Bayer, Hoechst, and Huls.

This habit of intraindustry collaboration in Europe and Japan stands in sharp contrast to the American experience. U.S. antitrust policy in general, and particularly the prohibition of price fixing under the Sherman Act, suppressed cartels and combines and instead encouraged large American firms to merge with one another when the need arose to coordinate strategic decisions. In France cartels had been less prevalent, not because of intense competition among members of an industry but rather because of long-established, tacit limits on competition, including a widespread view that price competition was unfair. In addition, because so many industries were oligopolistic, strong competitive forays would have only elicited destructive reprisals. In Great Britain, so-called gentlemen's agreements often made cartels unnecessary. Moreover, in all three European countries, industries without cartels often created combines that amalgamated and coordinated a large share of the industry's production.

Nevertheless, there remains a difference among the industry hierarchies in the four countries in which the VC problem was resolved through business-government cooperation. Although all four displayed a common pattern of strong industry associations, ownership linkages among the companies, and integrated VC-PVC producers, the Japanese industry lacked two features that made the European VC-PVC industries much more strongly hierarchical. First, there were far more companies in the Japanese industry; second, there was no evidence that large producers dominated the industries, or their industry associations, through market power, technical sophistication, extensive government contacts, financing of the industry association, or prominence on the national economic landscape.

Networks

The second set of institutional arrangements that distinguished the adversarial approach to the VC problem from the various forms of cooperation were the extensive, informal, middle-level networks of personal contacts that linked the business and government hierarchies in Japan and Europe and that were virtually absent in the United States. These networks blurred the boundaries between business and government in the four cooperative cases, while in the United States these parties dealt with each other at arm's length, across sharply defined organizational and legal boundaries.

The European quangos ("quasi-autonomous nongovernment organizations") were the clearest example of such networks. As subsequent chapters will show, these peculiar bodies merged many of the decisions and activities that companies, government agencies, and labor unions handled separately in the United States. In fact, in the three European countries, direct business-government relations were in large measure replaced by business-quango-government relations. In Great Britain, for example, neither industry nor labor dealt directly with the Ministry of Labor on the VC problem. Instead, their representatives worked together as members of a working group, itself a "mini-quango" established under the aegis of a government body called the Health and Safety Executive. The same pattern appeared in West Germany and France. Japan, however, stood alone in that business and government cooperated without relying on quangos. In part, this reflects the fact that the Japanese Ministry of Labor was created after

World War II by the U.S. Occupation and was modeled after U.S. institutions. Nevertheless, there was extensive, multipartite collaboration and shared responsibility for the development of the Japanese VC regulations, even though no formal quangolike body existed.

Company health and safety committees were another connecting link between the industry and government hierarchies in the four cooperative cases. In general, these committees had legal authority to gather company information, to investigate accidents, and to advise on, or sometimes make decisions on hiring and firing and on investments in capital equipment that affected workplace health and safety. In Great Britain and France, the final VC regulations required important features of the monitoring systems in plants to be approved by company health and safety committees. Moreover, quango and government inspectors were required by law to coordinate their activities with these company health and safety committees and to provide advice. In this way, government officials and officials of quangos had a direct, continuing influence on company decisions and operations relating to safety and health. They also shared some degree of responsibility for company practices, since the inspectors in the cooperative cases did not simply review a company's operations and then make a yes-no decision on whether the company was in compliance with the law. Instead, inspectors acted like technical advisers to company officials and their health and safety committees and helped them develop ways of achieving compliance.

During the 1970s, OSHA followed very different policies. In 1973 the Department of Labor had instructed OSHA "compliance officers" to issue citations for violations at a closing conference at the end of an inspection. On-site OSHA consultations would inevitably result in a citation if a violation was found, and a serious violation could result in the employer's being assessed a civil penalty as high as a thousand dollars per violation. All citations had to be posted permanently near the place of violation. Thus, OSHA inspectors were essentially policemen, and not, like their Japanese and European counterparts, intermediaries between industry and government.

Government ownership also blurred the boundaries among the major hierarchies in some of the cases of cooperation. In 1974 the Bank of England owned 20 percent of the shares of BP Chemicals and the government owned 48 percent. In West Germany, one of the largest VC and PVC producers was one-quarter owned by another West German company, VEBA, which itself was 40 percent owned by the West German government. Other connections also linked the government agencies and trade associations: in Japan, the executive director of the Japan PVC Association was usually a former official of the Ministry of International Trade and Industry. This was an instance of the Japanese practice of *amakudari*—literally, "the descent from heaven"—whereby senior government officials "retire" to positions in the industry partly to facilitate dealings with government. In the VC case, the Japan Development Bank also provided subsidized financing of industry efforts to reduce VC exposures. In Great Britain, officials from the Health and Safety Executive had been "on loan" for periods of a year to the Confederation of the British Industry where they served as chief health and safety officers. Finally, in Japan and the three European countries, chemical company officials have had relatively unrestricted, formal access to government officials. Conversation and consultations could take place in private, without public notice, and records and memoranda were not subject to anything like the American laws mandating "government in the sunshine."

In the United States, ongoing networks of personal relations spanning industry and government were extremely rare, especially for OSHA and the other new regulatory agencies. Instead, the agencies and the laws creating them sought to limit contacts between public and private officials to relations clearly defined by law, by contract, and occasionally by marketlike arrangements, such as the EPA's proposals to control pollution by auctioning "licenses to pollute." Part of the reason for restricting and controlling contacts between agencies and companies was concern that the new regulatory agencies not be "captured" by industry, as earlier agencies, such as the Interstate Commerce Commission, were believed to have been. The capture hypothesis held that older regulatory agencies had ceased to pursue the public interest and instead promoted the interests of the companies they were created to regulate. This was the result of a so-called life cycle of regulatory agencies. The gestation of an agency begins when an important and controversial problem galvanizes public outrage into a strong coalition supporting the creation of some new government body. The process of legislative review and compromise then dampens some of the urgency of that public outcry. Once the agency is established, public concern further wanes and the coalition disbands. After all, victory has been won, public concern moves to new issues, and media attention slackens. On the industry side, however, the battle has just begun. The newly regulated companies, having opposed the agency's creation in the first place, now turn their effort toward restraining and guiding its policies. Industry fights in court, member companies seek legislative restrictions on the agency, and company officials seek to build personal relations with their regulators to better influence agency policies. Regulators soon intuit that, if they ever leave government, they are extraordinarily well equipped to work as advisers for the companies they have been regulating. As a result, they moderate their policies and further accommodate industry interests. As the regulatory decisions become more detailed and technically complex, regulators rely even more heavily on industry experts. In the end, the agency has lost and forgotten its youthful crusading spirit and enters an enfeebled senescence as the ward of the industry it supposedly regulates.

The capture hypothesis became a near orthodoxy in the 1960s, and so the proponents of OSHA, the EPA, and the Consumer Product Safety Commission sought to make these new agencies capture-proof. Consequently, agency decisions and decision-making processes were opened to extensive judicial review, which a wide range of parties could initiate. The laws empowering these agencies were long and extremely detailed, especially in comparison with statutes creating the older, allegedly captive agencies. Detailed laws, and the congressional scrutiny such laws would permit, were considered a way of assuring that regulators would remain faithful to their original vows. Finally, the law required public hearings as part of the decision process. By opening the new agencies to such additional scrutiny, proponents hoped that agency officials would be less vulnerable to the seductions of regulated industries.

The desire of environmentalists and labor unions to avoid industry capture was part, but only part, of the reason for the creation of an OSHA that would stand at arm's length from industry. The early histories of the occupational health and safety institutions in the other countries suggest that, even if the capture hypothesis had not been politically and intellectually ascendant in the 1960s, OSHA would nevertheless have been created without networks linking it to companies. The creation of a new, large, and powerful agency requires the mobilization of great political force. The rhetoric of war and enemies, along with allegations of crisis and calamity, help mobilize large, active, and visible coalitions. The gathering of force on one side of an issue elicits countermobilization among opponents. The result is often a highly visible, acrimonious contest of charge and countercharge—circumstances unlikely to encourage the creation of an institution 'relying on networks of informal personal contacts. For example, the first French attempts to regulate the hours and conditions of work were regarded as:

> . . . an intolerable intrusion that could only undermine the authority of the master. To the requirements of Factory Act of 1841, the family enterprises of France, of northern France especially, opposed a deep, indignant immobility that discouraged examination in disarmed enforcement. The law called for voluntary inspectors among the manufacturers themselves, active and retired. It was a fiasco: few volunteered and many of these soon resigned in despair or under the pressures of friends and colleagues. There was no collaborating with evil.[6]

The early British legislation dealing with workplace health and workplace safety, and the inspectors who enforced it, met bitter opposition from employers and political leaders. A Scottish manufacturer described one of the early British Factory Acts as "indefensible in principle; invidious, oppressive, and absurd in its provisions; in its penalties harsh, ruinous, and tyrannical in the extreme." The National Association of Factory Occupiers started a petition drive to remove one of the first factory inspectors because of his efforts to enforce the legislation.

The struggle to create OSHA clearly fell within this pattern. At the 1968 Senate hearings on one of the bills that preceded the OSHA Act, Secretary of Labor W. Willard Wirtz testified that:

> . . . the clear central issue in S.2868 is simply whether the Congress is going to stop the carnage which continues because the people in this country . . . can't see the blood on the food they eat, on the things they buy, and on the services they get.

The April 1968 issue of *Nation's Business* rallied the opposition to OSHA with an article entitled "Life or Death for Your Business?" Its subtitle was "Labor Secretary Wants the Power to Shut You Down in the Name of Health and Safety," and the first paragraph read:

> Imagine yourself sitting in your office, a few months from today. A young man barges in. You recognized him as a man you once refused to hire. He had no education and no potential talent that you could use. His main experience consisted of cashing welfare checks.
>
> But he shows he is now a representative of the federal government—an "inspector" with the Department of Labor. And he threatens to padlock your plant and have you fined $1,000 a day if you don't do as he says.
>
> The young man—who knows nothing about your business—then tramps through your plant, without a warrant, ordering you to take costly steps to improve "health and safety."

An agency given birth through tumultuous conflict is unlikely to be structured or administered to promote cooperation through informal networks with the groups whose allegedly negligent or heinous behavior required the agency's creation. The first British factory inspectors were essentially policemen, like their OSHA counterparts 100 years later. They had magisterial power, which meant the inspectors themselves could try offenders and impose punishment. Similarly, OSHA's inspectors could not give advice or warning when they found a violation or a problem; the law required them to issue citations. And during the congressional debate on the OSHA Act, spokesmen for the Labor Department opposed any independent quangolike, multipartite body that would oversee the development and enforcement of standards, arguing it would resemble too closely the "cooperative" state programs that OSHA was intended to replace. The simple politics of good and evil, a precondition for the creation of powerful new agencies that disrupt existing patterns of power and influence, were reflected in relatively simple administrative arrangements designed to enforce laws and impose penalties. The much more complicated administrative arrangements, in which networks blur the boundaries between industry and government, tend to arise, if at all, many decades later in the lives of these agencies. It was roughly thirty years after the factory inspectors were first established in Great Britain that their roles shifted from pure enforcement to a mix of technical assistance and enforcing the law. The chief inspector, in 1878, attributed this change to the "increasing complexity of the work and the changing attitude of industry."

Moreover, antagonism among major parties persists after an agency is established, and the agency itself then becomes one of the contending parties. In 1973, for example, associates of Ralph Nader published a report on health and safety entitled *Bitter Wages*. The dominant metaphor of the book, quite literally, was that the blood of American workers was greasing the wheels of the American industrial juggernaut. The conflict over OSHA persisted because the new agency was implementing a wide range of practices and requirements that were an abrupt shift from previous cooperative business-state government relations based on consultation and the very limited use of penalties. Companies had to set up extensive new record-keeping systems; employees or their representatives could request inspections without the approval of the management; the Act granted workers a "bill of rights," and virtually all employers were covered by the OSHA Act's provisions. Adversarial thrust and parry—in the press, in Congress, and in court—dominated OSHA's first decade. The coalitions that had supported and fought creation of OSHA remained vigilant and active. To a significant degree, OSHA's first decade was

pervaded by adversarial activity simply because it was OSHA's
first decade.

The sharp boundaries separating American industry and
government in the VC case also reflected much broader historical patterns. For example, the large industrial corporation
came into existence in the United States *before* a large and
powerful central government hierarchy had been established,
while in the other four countries the order was reversed. As a
result, the U.S. federal government grew against resistance
from a relatively autonomous and powerful business community. Government action often ran directly counter to the
interests of powerful companies and powerful industries.
Moreover, the first extensive business-government relations
involving large American companies and government agencies produced highly publicized, corrupt cooperation during
the so-called Gilded Age. In the Credit Mobilier scandal, for
example, the Union Pacific Railroad bribed the vice president
and a number of influential congressmen. The Whiskey Ring
corruptly linked distillers with high officials of the Treasury
Department. Samuel Eliot Morison observed that ". . . these
industrialists, were not simply greedy (though most of them
were that); they fancied they were renewing Hamiltonian
policies, binding great financial and industrial interests to the
Federal government through mutual favors."[7] In fact, Morison sketches U.S. business-government relations a hundred
years ago in terms of powerful hierarchies linked by corrupt
and corrupting networks:

> The Federal Government was at the summit of a
> pyramid of corruption in the Northern States. . . . Jim
> Fisk and Jay Gould polluting the Erie Railroad by
> stock watering. . . . Collis P. Huntington buying the
> California legislature and bribing the Congressmen to
> promote transcontinental railroad interests; Peter
> Widener, obtaining street railway franchise by bribing

aldermen; John D. Rockefeller, using strong-arm
methods where chicanery failed to build his Standard
Oil empire. These were the conspicuous examples in
the middle tier of this indecent pyramid, the lower
courses of which were built by a sordid alliance between liquor, prostitution, and city police. . . .[8]

A different pattern appeared in the three European countries and Japan. All of these countries have had scandals involving industry officials and government agencies, but their
business-government relations had very different origins.
From the very beginning, elaborate networks linked industry
and government. Even before the Industrial Revolution,
strong central governments in Prussia, France, Great Britain,
and Japan had a direct hand in economic activity. The mer-
chant class that arose in Japan was closely tied to the territorial
lords who supervised, taxed, and when necessary, restrained
their activities. Under Colbert in the late seventeenth century,
the French government made extensive direct investments to
promote the clock, silk, lace, and carpet industries to secure a
steady supply of luxuries for the court. In both France and
Prussia, the government invested extensively in the small
manufacturers that produced armaments, and in Prussia the
monarchy was the largest producer of iron and coal in the
kingdom.

The British experience during the Industrial Revolution
was an exception to this broad pattern. Britain's industrial revolution occurred more or less spontaneously; the visible hand
of government had a very limited role. And precisely because
the British achieved extraordinary economic and military
power, both the French and Prussian, and later the German
governments became even more active in promoting and di-
recting economic activity. They did so through cartels, subsidies, state ownership, protection, and administrative guidance—all of which obscured the boundaries between the

public and private spheres. And after Japan was conquered by the four small boats of "barbarians" led by Admiral Perry, its political imperative became "rich nation, strong army," and its government adopted an industrial policy to gain economic and military might.

In the twentieth century, links between business and government grew stronger and more elaborate in the three European countries and Japan. Mobilization for two world wars, and the requirements of recovery from devastating losses, entailed decades of business-government collaboration. The rise of national planning in France and Britain provided another arena in which industrialists and government officials could exchange views and bargain over policies. And those two countries' nationalization of industry encouraged further interpenetration of the public and private sectors. Although the situation varied from country to country and from decade to decade, a broad historical pattern distinguishes Japan and the three European countries from the United States. The elaborate web of personal contacts linking industry and government—through quangos, health and safety committees, inspectorates, and other intermediaries—in the four countries that took a cooperative approach to the VC problem were merely small mosaics in a much larger, broader, and older pattern.

The case studies presented in the next chapters unfold against the background of these contrasting administrative arrangements for linking business and government. And once the five VC stories have been told, the broad framework of network and hierarchies provides a strong explanation of why the European and Japanese approaches to VC resembled each other so strongly—despite cultural and historical differences—and why their approaches all differed so greatly from that of the United States.

Chapter 3
THE UNITED STATES: THE ADVERSARIAL CASE

The U.S. approach to VC presents a classic case of adversarial business-government relations. The main contending parties were the companies in the VC-PVC industry, their industry association, OSHA, and the labor unions that represented VC and PVC workers. VC generated heated controversy from the time of the B.F. Goodrich announcement. The conflict intensified during OSHA hearings, continued during judicial review of the OSHA regulations, and diminished only after 1975 as companies achieved compliance with the new VC regulations and as the contending parties locked horns over other issues.

On 22 January 1974 B.F. Goodrich announced that within

the past two years, three workers from its Louisville PVC plant had died of angiosarcoma of the liver. VC and PVC companies responded swiftly. B.F. Goodrich, for example, immediately released all its available information on VC to workers in its VC and PVC plants; it advised all its customers and competitors of its findings; the company hired a consultant to study the records of all deceased Goodrich VC and PVC workers and told its engineers, scientists, and plant managers to reduce levels of VC exposure as quickly as possible—by purchasing equipment currently available and by developing new technology. This effort was coordinated by a group of executives and technical specialists who reported daily to the president of Goodrich's chemical division.

Parallel activities were started by other VC and PVC producers. Firestone arranged meetings between management representatives and small groups of workers at which all available information about the VC hazard was presented. It also began its own review of employee death certificates and work records. General Tire and Rubber formed a special VC team that included attorneys, engineers, and scientists. The group had its own jet and twice a month visited all fifteen sites where the company produced or fabricated PVC. Diamond Shamrock officials described a "great burst" of activity immediately after the Goodrich announcement, including purchases of protective and analytical equipment, and even the tearing down of walls at plants to permit VC dispersion.

OSHA also moved swiftly. On 30 January 1974 it published a notice in the *Federal Register* announcing a fact-finding hearing on VC for 15 February. The next day, NIOSH, OSHA's scientific arm, recommended a set of precautionary steps such as providing protective clothing and air-supplied respirators for the workers who cleaned reactor vessels.

The February fact-finding hearing was the first meeting of the major parties involved with the VC problem. These were

NIOSH and OSHA; large producers like B.F. Goodrich, Shell Chemical, and Dow Chemical; and officials of the United Rubber Workers, the Oil, Chemical and Atomic Workers, and the AFL-CIO. Professor Maltoni came from Italy and presented preliminary findings linking VC to cancer of the liver and other cancers in rats exposed to 250 ppm of VC.

On the morning of the hearing, B.F. Goodrich announced the death of a fifth employee from angiosarcoma of the liver. At the hearing itself, discussion centered on whether OSHA should immediately issue an emergency temporary standard (ETS) or proceed instead with its regular rule-making process. Under regular rule making, OSHA published a "Notice of Proposed Rule-Making" in the *Federal Register*. This described whatever regulation OSHA proposed for a particular problem, and it announced a period for public comment. OSHA also would have to hold a public hearing on a proposed regulation if an interested party so requested. On the other hand, if OSHA determined that workers were in "grave danger," it could issue an ETS that became effective immediately. Interested parties could still request a public hearing on an ETS, but the emergency procedures would obligate OSHA to issue its final regulation within six months. No such limit applied to ordinary rule making.

The rule-making issue provoked sharp conflict. The companies argued against an ETS, stressing that they had already reduced average VC levels below 50 ppm, a level at which no hazard had been demonstrated. The first signs of intraindustry conflict also appeared when some companies argued that OSHA's current VC standard of 500 ppm[9] should remain in effect, while others advocated an interim "working level" of 50 ppm. Union officials argued that VC was clearly a human carcinogen, that there was no evidence that any level of exposure to a carcinogen was safe, and that OSHA should issue an ETS requiring companies to reduce VC levels until it could no longer be detected, even with sophisticated instruments.

On 5 April 1974 OSHA issued an ETS: no employee could be exposed to more than 50 ppm of VC, and companies would have to monitor VC levels and evacuate employees or give them respirators when VC levels exceeded 50 ppm. OSHA said the ETS was necessary to protect workers from a "grave danger" of cancer. The agency chose the 50 ppm limit on the grounds that it was the lowest level industry could achieve immediately. OSHA also noted that several studies had failed to induce cancer in animals at levels below 50 ppm.

Under this emergency rule making, OSHA had six months to write a proposed standard, hold public hearings, make revisions, and publish a final standard. Consequently, during April, OSHA officials worked intensely on a proposed VC standard. A project officer and project attorney, under the supervision of OSHA's director of standards development, handled the task.

While OSHA drafted its standard, two further studies were published, both sponsored by the Manufacturing Chemists Association, a major trade association for the chemical industry.[10] In one, VC exposure induced angiosarcoma of the liver in rats exposed to 50 ppm. (These findings were announced just five days after OSHA issued its ETS for VC and defended an interim level of 50 ppm.) The other study was an epidemiological review of a large group of PVC workers. It found no cause of death with a statistically significant excess over what would have been expected in a comparable group of U.S. males. The researchers, however, were unable to locate all the workers they wanted to study.

By early May, just before OSHA announced its proposed standard, the worldwide death toll of angiosarcoma of the liver had reached nineteen. Thirteen of the victims were U.S. VC and PVC workers.

The "Bombshell": OSHA's Proposed VC Standard

On 10 May OSHA published its "Proposed Standard for VC." VC exposures were to be set at the "no detectable level, as determined by a sampling and analytical method capable of detecting vinyl chloride at concentration levels of one ppm." Companies would be required to comply with the regulation through engineering changes and work practices (and not by requiring workers to wear respirators), and they had to comply "as soon as feasible." OSHA also announced that it would hold public hearings on its proposal in Washington, D.C., starting on 25 June.

The "zero-acceptable" ppm level was the lightning rod for industry objections. As one company official put it:

> That requirement made a lot of people in the industry furious. I mean mad enough to start fist fights. People said, "They can't do that to us—we're running a clean operation." When further OSHA announcements came out, the attitude was, "Did you see what those bastards did to us today?"

The shock of OSHA's proposal prompted the VC and PVC companies to seek a common position for the 25 June public hearing. Members of the Society of the Plastic Industry (SPI), an association representing most of the major plastics producers in the United States, organized these efforts. Before OSHA had announced its proposed regulation, many SPI members were reluctant to have the association become involved with the VC problem as it had little experience dealing with OSHA on health matters involving chemicals. (The Manufacturing Chemists Association had generally handled these problems.) The SPI had many other responsibilities to the vast majority of its members who were neither VC nor PVC producers, and active involvement in the VC issue raised the risk of adverse publicity for all SPI members. But this reluctance vanished when OSHA proposed the "no detectable" limit. Even companies not producing VC or PVC wanted the SPI to take an active role, in part because some of them feared they could be next. In response, the SPI arranged the first of what proved to be a long series of intraindustry meetings.

Approximately sixty representatives of VC and PVC producers attended the first meeting. They came from twenty-four SPI member companies and from two nonmember companies. Most companies sent an executive from their VC or PVC operations, an attorney, and a scientist or engineer, familiar with some aspect of the problem. The group of sixty was too large for detailed discussions, so several task forces and an overall coordinating committee, representing the major producers, were established. These smaller groups held frequent discussions during May.

The meetings culminated in early June at another plenary session whose objective was agreement on a common industry position. All the companies agreed that it was not technologically feasible for them to comply with OSHA's proposed exposure limit under the current manufacturing procedures. The question, therefore, was what exposure levels might become feasible over the next few years. This elicited long, heated discussions. At one point, Anton Vittone, the president of B.F. Goodrich Chemical, threatened to resign as chairman of the meeting because he disagreed strongly with representatives of several other major producers who wanted to recommend a permanent standard of 50 ppm. Vittone believed that levels near or even below 10 ppm could be reached within a few years. Further discussion produced a compromise: the SPI would recommend an average exposure level of 25 ppm, to be effective in October 1974, and a level of 10 ppm, to be effective in October 1976. However, companies that disagreed with this timetable were free to testify on their own at the OSHA hearing.

During May and June, scientific groups and labor unions

had also been active. The New York Academy of Science held a meeting of 300 scientists from eight countries to discuss the VC hazard. They heard evidence of blood, liver, and respiratory abnormalities, as well as reduction in liver function, enlargement of the liver and spleen, and excess production of red blood cells—all apparently related to VC exposure. Two additional angiosarcoma deaths were reported, and these were especially disturbing. One victim was a General Electric employee who had worked with PVC insulation; the other was a woman who had lived for years downwind from a PVC plant. These reports received widespread attention because they suggested that VC could harm people who did not work in VC or PVC operations.

During the spring, several union leaders took public positions on the VC problem. Anthony Mazzochi of the Oil, Chemical and Atomic Workers, which represented about 1,200 VC and PVC workers, said that "workers are tired of being used as guinea pigs" and called for a VC exposure limit of zero. Peter Bommarito, president of the United Rubber Workers, sent a message to President Nixon, several senators, and other officials. It read:

> Deaths from angiosarcoma of the liver and other organs are now reaching epidemic proportions not only in this country but also the world. Delayed and capricious action . . . cannot be tolerated. . . . withholding data has already cost the lives of countless workers involved in the production and use of polyvinyl chloride resin. Stop the cruel and unnecessary killing and injuring of our work force. [11]

In public statements, industry officials responded by calling the proposed standard "impractical," "unachievable," and "a disaster for the country." A spokesman for Firestone said the standard ". . . would literally cripple the industry." And the SPI released a study by Arthur D. Little, Inc., which concluded that a shutdown of VC-PVC production would cost 1.6 million jobs and reduce the gross national product by at least $65 billion.[12]

Such industry statements provoked further comments from union leaders. Sheldon Samuels, director of occupational health, safety, and environmental affairs for the AFL-CIO, called the Arthur D. Little study "meaningless" and said the union had information indicating that industry could meet the proposed standard. Peter Bommarito of the United Rubber Workers accused industry of "trying to scare everybody about losing a job."

For its part, OSHA officials sought to avoid the fray. Dr. Daniel Boyd, who was OSHA's director of standards development and was supervising preparation of the VC regulation, responded to industry's charges by acknowledging that the proposed standard would be "tough to achieve," but emphasized that it was flexible and allowed companies to use a combination of engineering controls and work practices to meet the exposure limit.

All the activities that preceded the OSHA hearing—the medical reports, the rising death toll, and the public positions taken by the company officials, the SPI, and union leaders—were covered extensively by the media. In May, *Time* magazine ran the first of two articles on VC. It quoted a NIOSH official who said, "I would suspect that this [angiosarcoma of the liver] is going to be the occupational disease of the century."

The stage was now set for the OSHA hearings. Reviewing the last few months, one OSHA official said he expected the hearings to be "the biggest show we've ever had."

The OSHA Hearings

OSHA's aim at the hearing was to gather information that would help in the drafting of a final VC standard. General criteria for such standards are given in the OSHA Act: for toxic substances, OSHA must adopt the standard

> . . . which most adequately assures, to the extent feasible, on the basis of the best available evidence, that no employee will suffer material impairment of health or functional capacity even if such employee has regular exposure to the hazard dealt with by such standard for the period of his working life. . . . In addition to the attainment of the highest degree of health and safety protection for the employee, other considerations shall be the latest available scientific data in the field [and], the feasibility of the standards. . . .

The public hearings lasted eight days, with the official record held open until August so parties could submit further information. In the end, the VC record exceeded 4,000 pages and included over 800 oral and written submissions. The Department of Labor, industry, labor, OSHA, other government agencies, public interest groups, independent experts, physicians, and specialists from a wide range of fields, all presented testimony and cross-examined each other before an administrative law judge. Media coverage was intense.

Two issues predominated: whether the medical evidence justified the "no detectable" limit, and whether it was technologically and economically feasible for industry to comply with the limit.

On the medical issue, industry spokesmen argued that current angiosarcoma cases were the result of the very high exposures that had prevailed in the industry ten or twenty years earlier. Major VC and PVC producers—including Diamond Shamrock, Dow Chemical, B.F. Goodrich, Tenneco Chemicals, Union Carbide, and Firestone—testified that extensive surveys of their employees, including those exposed to VC more than ten years earlier, had revealed no further deaths from liver disease.

Several prominent witnesses countered the industry position.[13] Dr. Irving Selikoff, professor of medicine at Mount Sinai School of Medicine, cited a study of 400 Goodyear PVC workers that found obstructed lung infection in 58 percent of the workers and indications of brain and lymphatic cancer. Selikoff argued that the vinyl chloride industry ". . . is a relatively young one, and the late effects of toxic exposures have only begun to appear; most of our experience is ahead of us." Dr. John Peters of the Harvard School of Public Health cited an ongoing study of the records of 161 deceased VC and PVC workers. It indicated a 50 percent excess of deaths due to cancer, including a 320 percent excess of brain cancers and 1,000 percent excess of liver cancer. Cancer specialists from NIOSH and the National Cancer Institute testified that no threshold level had been established for human carcinogens. The AFL-CIO supported the proposed standard with the qualification that "substantial risks" remained even at levels as low as 1 ppm.

On the issue of technological feasibility, a long series of company officials stated flatly that industry simply could not achieve the exposure level required by the proposed standard, and that attempts to do so would cripple the industry. Industry's principal objection was to the "no detectable" ceiling: one witness stressed the point by saying that even production systems that "cannot" leak inevitably do leak, and hence a company with a "leakproof" system could find itself in violation of OSHA's standard. Furthermore, the proposed standard required workers to wear bulky air-supplied respirators if exposure limits were not being met, and both industry and labor officials argued that these were extremely uncomfortable for workers and could even produce their own safety hazards. Industry witnesses also objected to the inclusion of PVC fabricators under the standard because residual VC levels in PVC resins were already below the proposed exposure levels; therefore, they argued, monitoring and record keeping in fabrication plants were unnecessary. In addition to these major objections, industry raised many lesser but nonetheless significant questions. For example, Union Carbide pointed out that the proposed standard was written ambiguously and could be interpreted to mean either that (a) a company with VC concentrations of nearly 3 ppm would be in compliance, or (b) that a company maintaining levels as low as .5 ppm would be out of compliance. Industry also objected to the "no detectable" language on the grounds that companies might be compelled to reduce exposure levels indefinitely as new generations of increasingly sensitive detection equipment became available.

The industry positions were supported by massive documentation. The Firestone statement was 156 pages long, had a 222-page appendix, and weighed over four pounds.

Industry spokesmen also proposed alternatives to the OSHA standard. The president of SPI advocated a compromise based on a phased reduction of VC levels. Some firms had accepted this proposal, but others expressed skepticism. Air Products, for example, stated that levels below 25 ppm would be "much more difficult and more speculative as to achievements." Firestone's testimony stated that its engineering studies

> indicate that it is impossible on the basis of the present state of technology and the engineering art to predict with any degree of certainty that the reduced level SPI suggests can be achieved in existing facilities.[14]

On the other hand, B.F. Goodrich stated that it had already achieved average exposure levels between 12 and 14 ppm, and the Calgon Corporation testified that technology already existed for reducing VC levels as low as 1 ppm.

The issue of economic feasibility centered on estimates of the cost of achieving various VC levels in the workplace. Firestone had commissioned a study by an independent design engineering firm. It concluded that, for Firestone alone, a 50 ppm limit would cost $10 million; a 25 ppm average, $20 million; a 10 ppm average, $40 million "with no assurance that such an exposure level could be achieved by technology." Zero to 1 ppm was estimated to cost more than $55 million, but the study concluded this level could not be achieved.[15]

OSHA had commissioned its own study, but it was not completed until late August. The study—prepared by Foster D. Snell, an economic consulting firm—estimated capital costs for the entire VC and PVC industry. Estimates ranged from $50 million for a 50 ppm ceiling to $229 million for a 10–15 ppm level. The study relied on the Firestone estimate of the cost of a 0–1 ppm standard and extrapolated total costs for the industry of nearly $900 million.[16] Firestone challenged some of the assumptions of the OSHA study as optimistic, except, of course, for the 0–1 ppm data. Supporters of OSHA's proposal argued that the OSHA study was unduly pessimistic and called industry's argument "blackmail." The Health Research Group, a public interest group affiliated with Ralph Nader, testified:

> The economic impact prepared for the Department of Labor, the Snell study, commits the same fundamental error as an earlier study for the vinyl chloride industry prepared by the Arthur D. Little Company. It accepts unquestioningly the industry assumption that it is technologically infeasible to comply with a standard of "no detectable level" of VCM. This irresponsible action is tantamount to assuming the virtual cessation of the industry as the industry contractor did. . . . Moreover, this unthinking assumption incorporates wholecloth the incessant chorus of industry witnesses parroting the claim of industry unsubstantiated by any probative evidence.[17]

After the hearings, OSHA had ten weeks to review the hearing records and write a final VC standard. During this period rumors abounded. One rumor suggested that OSHA would set a compromise standard of approximately 25 ppm in hopes of avoiding lawsuits from either industry or labor. According to another rumor, OSHA technical personnel favored a limit of 10–20 ppm, while more politically minded officials wanted levels in the low, single digits.

The Final OSHA Standard

OSHA published its final VC standard in October 1974. Its key provision was:

> The standard sets an exposure limit of 1 ppm averaged over any eight hour period, and a ceiling of 5 ppm averaged over any period not exceeding 15 minutes.

The standard would become effective 1 January 1975. Until then the ETS of 50 ppm would be in effect. Companies had to achieve 1 ppm through engineering controls and work practices and not by using respirators. Companies were required to monitor workplace VC levels and to make all monitoring results available to workers.

Within *two minutes* of OSHA's announcement, the SPI, Hooker, Union Carbide, Firestone, Tenneco, Air Products, and B.F. Goodrich had all filed suits in various circuit courts. These petitions were filed under provisions of Section Six of the OSHA Act that allow "any person adversely affected by a standard" to challenge it in a circuit court. In these cases, the court's mandate was to determine whether OSHA's decision was "supported by substantial evidence in the record taken as a whole." All the industry petitions were consolidated for review by the Second Circuit Court of Appeals in New York in a case named *Society of the Plastics Industry, Inc. v. OSHA*. The AFL-CIO intervened on behalf of OSHA. At the start of the proceedings, the court stayed the effective date of the standard until its review was complete.

In court, industry made the same two arguments it had in the hearings: the absence of medical justification for a 1 ppm standard, and infeasibility. Attorneys for OSHA and the labor unions also responded with data and arguments they had presented at the VC hearings. In January 1975 the court ruled in favor of the OSHA standard.

The appeals court acknowledged that the VC hazard was "on the frontiers of scientific knowledge," but went on to state that OSHA had a duty to act to protect the worker, even when "existing methodology or research is deficient." The court also recounted the evidence from the VC hearings, which it considered sufficient to justify the OSHA standard.

On the feasibility issue, industry attorneys had cited the virtually unanimous testimony by company officials that existing technology could not meet such a standard. The court's response to this argument was that the hearing record did not show that the 1 ppm standard was "clearly impossible" to achieve, and the judges' opinion pointed to evidence in the hearing record that much progress had been made already and that many "useful" techniques were yet to be implemented. The court also said that OSHA could set standards that went beyond the technological status quo and suggested that the VC and PVC manufacturers simply needed "more faith in their own technological potentialities." The court also commented on its own task of reviewing the VC hearings record: "The examination of the 4,000 page record in this case has been a prodigious task, aggravated by duplications of testimony, irrelevant exhibits and letters, almost illegible reproduction of documents, and a generally blunderbuss approach in the petitioner's briefs."

Industry appealed to the Supreme Court, but in May the Court refused to hear the case. The SPI then acknowledged that all legal channels had been exhausted and said that the industry would now "concentrate its resources on compliance and, where appropriate, work to bring about reasonable modifications, interpretation, and administration of the standard."[18]

Compliance: Did Industry Cry Wolf?

By April 1976, the VC and PVC industries had generally complied with the OSHA standard. OSHA tests in 1976 and 1977 found that 90 percent of the samples taken were in compliance with the standard, though there were temporary exposures as high as 25 ppm. This did not mean, however, that all companies had achieved 1 ppm solely through engineering controls and work practices. Some cases required extensive use of respirators. The standard permitted this, but only as an interim measure. What happened in many cases, particularly with older plants, was that engineering controls reduced VC levels to between 1 and 5 ppm. Compliance was then achieved because workers wore respirators and because they did not spend an entire shift in exposed areas, and thus their personal, time-weighted exposure averaged below 1 ppm.

The total cost of complying with the standard has been estimated at $200–280 million.[19] This included capital costs of approximately $130 million, annual operating costs of $7–10 million, whose present value (at a 10 percent discount rate) is $70–100 million, and $30–50 million to replace capacity lost as a result of closing down capacity. B.F. Goodrich alone spent nearly $35 million in the largest research, development, and engineering project in its history. Three companies shut down facilities rather than operate them under the new regulation. Some of these plants, however, were old and inefficient, and the VC standard was actually the last straw for them, so the full cost of lost capacity cannot be assigned to the standard. Approximately 375 jobs were lost.

Industry's swift compliance raised a further issue that the *New York Times* summarized in the headline "Did Industry Cry Wolf?" The industry defense was that it had complied with the *final* VC standard, which it claimed was more lenient that OSHA's *proposed* standard. However, critics of the industry, as well as some OSHA officials, described this distinction as "just a little face-saving semantics." Critics also said that the industry had exaggerated the costs of complying with the 1 ppm standard. They noted that Firestone had projected capital costs of $40 million for achieving a 10 ppm standard at just two of the plants. Yet total capital costs of achieving 1 ppm throughout the industry were approximately $150 million.[20]

In any event, by late 1976, the problem of workplace exposure to VC in the United States was basically settled. In September 1976 a *Chemical Week* article was appropriately entitled "PVC Rolls Out of Jeopardy Into Jubilation."

Chapter 4
GREAT BRITAIN, WEST GERMANY, AND FRANCE: COOPERATION THROUGH QUANGOS

The approach to the VC problem in Great Britain, France, and West Germany stands apart from both the U.S. and Japanese efforts. Unlike their American counterparts, the European industry and government officials worked together—in small VC working groups—to resolve the VC problem. They cooperated over several years, across a wide range of major and minor issues, and they did so through elaborate and quite unusual institutional arrangements. It is precisely these arrangements that distinguish the European cases from the Japanese. Business and government also cooperated in Japan but did so without relying, as the Europeans did, on quangos ("quasi-autonomous nongovernment organizations"). The name "quango" is apparently British and has spawned a small family of neologisms such as the noun "enquangment."

In the three European countries, quangos involved with the VC problem took one of two forms. One type was the British Health and Safety Commission (HSC). This body has official responsibility for overseeing workplace health and safety in Britain. Its nine members are appointed by the secretary of state for employment, a cabinet minister, after consultation with representatives of employers, employees, local government, and professional associations. In practice, the Confederation of British Industry and the Trades Union Congress (TUC)—the dominant umbrella organizations for Britain industry and labor—have been highly influential in the choice of members. In 1977, for example, the HSC included the head of the TUC's Social Insurance and Industrial Welfare Department, an official of the Union of Construction, Allied Trade and Technicians, and the deputy chairman of the plastics division of Imperial Chemical Industries. HSC policies are administered by its operational arm, the Health and Safety Executive (HSE). In 1977 the HSE had a staff of approximately 4,000 and a budget of approximately $100 million. Staff activities included information and advisory services, research, and inspection of workplaces by the Factory Inspectorate. The HSE also established special working parties—with members drawn from unions, companies, the HSE itself, and other bodies—to develop guidelines and codes of practice for particular occupational hazards. One of these working groups was responsible for resolving the VC problem in Great Britain.

The other form of quango appears in France and West Germany and differs from the British quango in two principal respects. First, it is managed by boards of directors that represent only labor and management, and not other parties. Second, such quangos are just one-half of a *dual system* for supervising workplace health and safety. That is, the quangos had their own inspectorates, their own research staffs, their own budgets (provided by contributions from employers), and developed and enforced their own guidelines and recommendations. At the same time, the Ministries of Labor in both countries have separate inspectorates, research facilities, and systems of regulation and enforcement. The two halves of the dual system attempt to coordinate their efforts, and the Ministries of Labor supervise the quangos.

The West German quango with responsibility for the VC problem was the Industrial Injuries Insurance Institute of the Chemical Industry (*Berufsgenossenschaft der Chemischen Industrie*, or BG Chemie). The French quango was the National Sickness Insurance Fund (*Caisse Nationale de l'Assurance Maladie*, or CN). The French regulations for vinyl chloride were developed by a National Technical Committee for the Chemical Industry, part of the CN, consisting of nine employer and nine employee representatives.

All three quangos were large institutions. The British Health and Safety Commission, the West German BG Chemie, and the French CN each had their own funding, their own managers and employees, and clearly defined responsibilities. The three bodies were governmental because they were closely supervised by government officials and because they performed functions that, in other countries, are performed exclusively by government agencies. Nonetheless, these bodies were quasi-autonomous: their basic policies and ongoing management were the responsibility of representatives of industry and labor.

In the European countries, as in the United States, VC levels in the workplace were very high in the 1950s and 1960s. Estimates generally place these levels at approximately 1,000 ppm in the late 1940s and early 1950s, at approximately 500 ppm in the late 1950s, and at 300–400 ppm during the

1960s.[21] In all three countries, the B.F. Goodrich announcement in January 1974 immediately triggered hectic activity by VC and PVC producing companies to lower VC exposure levels as quickly as possible, as well as reactions by trade unions and government.

Great Britain

British authorities and company officials learned about the B.F. Goodrich announcement almost as soon as their American counterparts because the Manufacturing Chemists Association contacted the two major British VC and PVC producers, British Petroleum Chemicals and Imperial Chemical Industries (ICI), immediately after Goodrich issued its press release. The companies then passed the information on to the chief inspector of factories.

Within a week of the Goodrich announcement, ICI had given government departments, the Trades Union Congress (TUC), and its own workers and customers all the information on VC it had available. ICI also arranged a private discussion between some of its senior officials and scientists and the TUC medical adviser. After this meeting, the company held management-worker meetings at which trade union health officials, the TUC medical adviser, members of the Factory Inspectorate (a part of the Health and Safety Executive, or HSE) discussed the Goodrich announcement and answered questions. BP Chemicals and the other two British PVC producers took similar steps.

The companies also began meeting with officials of the HSE. ICI and BP Chemicals took the leading role, since they were by far the largest producers and had the strongest technical and scientific staffs. The meetings brought together scientists and senior HSE officials responsible for occupational health. At the early meetings, officials from the Factory Inspectorate stressed that the companies should urgently take all possible steps to reduce VC exposure. The officials also arranged for the mobile laboratories of the Factory Inspectorate—the "shock troops," in the words of one official—to visit all VC and PVC plants to review company efforts. The companies also agreed on an initial voluntary objective of reducing VC levels below 50 ppm, and the parties also agreed to start epidemiological reviews of VC and PVC workers.

This network of contacts among government, industry, and labor union officials remained in place through the entire British effort to reduce VC exposures. In fact, the VC working group (discussed below), which was established under the HSE and which ultimately resolved the VC problem, included many of the individuals involved in these early meetings. Industry, government, and labor officials communicated with each other at meetings, by memo and letter, and by telephone. The meetings were held in private and were subject to one self-imposed constraint: bipartite discussions of major issues were to be generally avoided. The tacit understanding among the parties was that important matters would be discussed only when all three parties were present.

During the winter and early spring of 1974 another important set of relationships evolved. These were the intraindustry arrangements set up under the auspices of the Chemical Industry Association (CIA), the principal trade association for the British chemical industry and the normal vehicle for such collaboration. The joint effort allowed companies to pool their engineering, industrial hygiene, and other specialized skills. Companies could also concentrate on technical problems, while the CIA handled public relations and the media. A final reason for the CIA arrangement was that the association had long-established working relations with the government agencies dealing with the chemical industry.

For the next two years, a VC committee of the CIA was the forum in which companies coordinated their efforts to eliminate the VC hazard. Information from companies was also communicated directly to the HSE's VC working group by written reports and by individuals who were members of both the CIA's VC committee and the HSE's VC working group. For example, A.W. Barnes, a director of the ICI Plastics Division, was the head of the CIA's VC committee and the member of the VC working party that developed the British Code of Practice for VC. In fact, Barnes worked virtually full time for nearly two years on the VC problem.

Later in the spring, HSE's Factory Inspectorate invited industry and labor to form a joint working group for VC that would prepare a detailed code of practice for VC. The TUC nominated the labor union members of the working group, after consultation with the labor unions that represented VC and PVC workers. (As the British umbrella organization for labor unions, TUC has no direct control over member unions; it offers mainly a central forum for discussions as well as services such as arbitration and conciliation.)

Industry members were nominated by the Confederation of British Industry (CBI), whose membership included private companies, public corporations and nationalized industries, and employees' organizations and trade associations. The CIA, for example, nominated representatives to the Council of the CBI as well as to the specialized committees through which many CBI activities were carried out. Approximately 10,000,000 British workers were employed by companies associated with the CBI. The CBI's activities included highly publicized meetings between its leadership and the prime minister or chancellor of the Exchequer, private briefings of civil servants and members of Parliament during the early stages of legislation, public relations efforts on television and radio, and irregular but frequent discussions with the TUC. Although the CBI was formally responsible for nominating industry members of the VC working group, it did so upon the advice of the CIA and the VC-PVC producers, since they were directly familiar with the issues that would form the working group's agenda.

This VC working group first met on 14 June 1974 and was chaired by the chief inspector of factories. There were five labor union representatives, two from the TUC headquarters (including the TUC medical adviser), a TUC health and safety officer, and two officers of individual labor unions. The five industry representatives included an official of the Confederation of British Industry, Barnes from ICI, an ICI occupational physician, and a senior manager from the plastics operations of BP Chemical. Five officials from the Factory Inspectorate and the Employment Medical Advisory Services represented the government. Another group of government officials attended the meetings as a secretariat. It included, among others, a medical statistician, the deputy chief of the Alkali Inspectorate (which was responsible for air pollution problems), and representatives of the Department of Health and Social Services.

At this first meeting, industry officials reported that their companies had already reduced VC levels below 50 ppm, a considerable drop from the 200 ppm level the government had required since 1972. The working group agreed, after discussion, that industry should next aim for a 25 ppm level, although the labor representatives agreed only on the understanding that 25 ppm was an interim measure and that further investigations would examine the feasibility of lower levels. The group also agreed that its next target would be 10 ppm, once appropriate equipment was available.

The other important decision at the first meeting was to establish two subgroups. Each would be chaired by a senior government official. A medical group would gather information on the VC hazard, oversee epidemiological efforts, and draft the parts of a code of practice covering medical examinations. The other group would draft the rest of the code that addressed issues such as monitoring and control, computerization, reactor cleaning, protective clothing, and so forth. Each subgroup had six members: two from the labor union members of the main working group, two from the CBI representatives, and two from the government side. Like the main working group, the subgroups called on outside experts who attended meetings at which specialized technical matters were discussed. Between June and December 1974, both subgroups met frequently—generally, at least once every two weeks— and the meetings often lasted an entire day.

Altogether, the main working party met seven times between June 1974 and May 1976. During this time, the size and composition of the group changed. Two more industry representatives were added, so that all four VC and PVC producers were represented, and these were balanced by two more labor union representatives. Some industry representatives were later replaced by others, and representatives of other government departments—such as the Departments of Trade, Health, and Environment—attended several meetings, as did various specialists from the HSE. All the meetings were conducted in private in government offices. The working group issued no statements or reports until February 1975, when it published a "VC Code of Practice." However, all parties did inform the organizations they represented about the discussions and decisions of the working group.

Other meetings also took place under the auspices of the main working group and its subgroups. For example, when the draft code of practice was being written, one or two government officials would meet from time to time with industry specialists to discuss particular questions and draft portions of the code. These drafts were later reviewed, word for word, by the full subgroup and by outside groups, like the CIA's VC committee.

Between June and December 1974, while each subgroup worked on its special tasks, the main working group met only four times. The meetings lasted about half a day, except for the final day-long meeting at which the VC Code of Practice was discussed and approved. The role of the main working group was to disseminate accurate information among the parties, to discuss progress reports from the subgroups, and to review the results of animal studies, epidemiological studies, and technical studies from around the world. At each meeting, industry representatives and members of the Factory Inspectorate reported on the progress in reducing VC levels. Labor union representatives reported on the concerns of their members, discussed information they had received from their counterparts in other countries (including reports on OSHA's activities), and raised a number of shop floor problems such as the reluctance of most workers to wear respirators for extended periods.

At its December 1974 meeting, the main working group reviewed line by line and approved a draft code of practice. This was published two months later in a "temporary format" so that subsequent changes could be incorporated. In the code, the "interim hygiene standard" for VC was defined by:

> . . . a ceiling value of 50 parts per million, that must not be exceeded; the average exposure over the whole shift must not exceed 25 parts per million. These are outside limits and it is further asked that exposure should be brought as near as practicable to zero. This standard, which is to be kept under review by the Working Group, therefore encourages progressive reductions of exposure to the lowest possible levels given present engineering knowledge.[22]

The code required the monitoring of VC levels at least once per shift. It provided guidelines for monitoring and described the information that a company's monitoring plan should contain. In general, however, matters such as the frequency of measurement and the type of equipment were left to the discretion of the company. The code said that a senior manager should be responsible for assuring that there was "joint consultation with employees' representatives" on all aspects of the VC problem, including the results of the monitoring. Workers were required to wear respirators when VC levels exceeded the hygiene standard, but the code emphasized that the companies should aim "to provide plant and equipment so designed as to assure maximum containment of vinyl chloride." The code was thirty-nine pages long and included eleven appendixes and eight pages of sample forms for recording monitoring results.

During 1975 the main working group met just twice. In April the group reviewed a report from the medical subgroup on recent studies of the effects of VC exposure and on British cases of angiosarcoma of the liver. By this time there were two British cases with confirmed links to VC and a third with a suspected link. The group also heard reports from industry and from the Factory Inspectorate on the code's implementation. On the basis of these reports, the group decided not to change the 25 ppm hygiene limit. It also decided that six of its members, two from each side, would visit all six VC and PVC plants in Great Britain and review industry's efforts to reduce VC.

The visits took place over the next several months. During each, the group discussed the temporary code of practice with workers and answered their questions about the VC hazard and about the monitoring and control technology. The group also reviewed the new equipment, training material, emergency procedures, and other aspects of a company's compliance efforts.

The plant visits also provided group members with opportunities for additional discussion of the problem and for further development of personal relations. The members had become acquainted during the working group meetings and in the informal discussions. A few of the members—mainly on the government and trade union sides—had known each other before the VC episode. However, the trips to plant sites encouraged especially strong relationships. They said they had "no holds barred discussion" and much pleasant "chit chat." Several members of the working party thought that the informal discussions and personal acquaintance enhanced the effectiveness of the working group. One of the members later commented that "credibility is the key to the tripartite equation" and added that the extended personal contacts among working group members were essential in developing mutual respect and confidence. Some of the trade union members acknowledged that they had entered the working party somewhat warily, that the industry and labor sides at first "eyed each other up." Later, however, after many discussions and after observing the industry efforts to control VC, the parties achieved a "real breakthrough in relations" and got along "famously." Other participants described the working group process as one of "forging new relations" and industry members described their trade union counterparts as "level-headed and responsible men."

The second working group meeting in 1975 took place in October. The group reviewed the monitoring plants for all six companies and amended the hygiene standard to read:

> The interim hygiene standard for vinyl chloride is as follows: ceiling value of 30 parts per million with a time-weighted average of 10 parts per million, allowing that wherever practicable exposure should be brought as near as possible to zero concentrations.[23]

The final working group meeting took place in May 1976. The group approved some minor changes in the code and then sent it to the Health and Safety Commission for review. After the meeting, the VC working group became a subcommittee of the HSC's Advisory Committee on Toxic Substances. Subsequent meetings did not, however, lead to changes in the hygiene standard because companies had already achieved exposure levels well below 10 ppm. This was also true of the first recommended limit of 25 ppm: at the time it was promulgated, companies were already in compliance. In fact, by the time the final standard was published, exposure levels were in the very low single digits. In the summer of 1975, average weekly VC levels in British plants were somewhere below 5 ppm, and some plants had reached 2 ppm. By 1977 plants in Great Britain were maintaining levels between .5 and 3 ppm. ICI's capital costs for complying with the VC Code were approximately £10–15 million, which implies capital costs for the whole British VC-PVC industry of over £20 million.

Compliance was achieved by what the deputy chief inspector of factories called "Herculean efforts" by the producing companies. In practice, this meant that the companies, acting through special VC "project teams," had taken approximately the same steps to reduce exposures as VC and PVC producers in other countries. They "plugged leaks," installed automated systems for cleaning polymerization vats, and so forth. Exposure levels were monitored by similar systems in the four British companies. These measured VC levels at a number of fixed points in the workplace as often as once per minute. As a result, nearly 500 readings could be taken over each shift. The producing companies installed these systems as a result of discussions and informal agreement within the working group. The code itself required only one sample per shift, but there was a clear understanding that computerized multipoint systems would actually be installed in the plants.

The final VC code involved a careful balancing of costs and risks. Such balancing had been mandated by a British court decision in 1943. The focal point of the cost-risk issue was what several working group members called the "numbers game": the question of the appropriate levels—ceiling and average—for VC. This was a lightning rod for debate and negotiations on this point throughout the life of the working party. Several members of the working group stated that this "game" was played reasonably and fairly, and they gave several reasons for this. One was the agreement that accompanied the first hygiene standard that "exposures should be brought as near as practicable to zero." Another was the attitude of trust and mutual confidence that emerged among working group members. This meant that the judgments of company scientists and managers on technical feasibility and costs were taken seriously. Nevertheless, the labor union participants stressed that they had independent ways of checking what industry representatives said. One of the labor representatives commented:

> We can rely on the inspectors of the Health and Safety Executive to check on industry. We have contacts with trade unions overseas and learn what companies are doing elsewhere. We have affiliate unions with technically skilled members who can evaluate industry claims if need be.

There were no judicial appeals of the VC Code of Practice. One reason is that the working group members had already agreed on all the key issues—hygiene limits, monitoring, joint consultation, and so forth—and industry had complied fully with the terms of the code. Furthermore, neither companies nor trade unions could actually seek judicial review. In general, British health and safety regulations cannot be reviewed in court until a person or company has been charged with a violation. The VC Code of Practice was further sheltered from court challenge by its peculiar legal status. It was not even an *approved* code of practice, as defined in the Health and Safety at Work Act of 1974. It was simply a voluntary agreement, among members of the working party, with which industry had complied.

West Germany

From an American viewpoint, the West German institutional arrangements were even more striking than the British because of the German dual system for occupational health and safety. In addition to relying on a quango, BG Chemie, the Germans also used two separate multipartite working parties—acting in tandem—to resolve the VC problem. One working party was organized by the Ministry of Labor and developed legally enforceable exposure limits for VC, the rough legal equivalent of the U.S. VC regulations. The other working party, organized by BG Chemie, the West German variant of the British Health and Safety Commission, developed guidelines suggesting ways in which companies could comply with the regulations of the Ministry of Labor. These guidelines were the rough equivalent of the British Code of Practice for VC.

BG Chemie merits introduction because of the important role it played in the VC problem and because it is such an unusual institution. BG Chemie is one of thirty-five industrial injuries insurance institutes in West Germany. Each BG is responsible for one or more major industrial sectors, and every employer in a sector must belong to the appropriate BG; all workers in a sector are covered by its BG. The BGs have two principal roles: they pay compensation to injured workers, and they help employers prevent injuries to workers. Each BG is autonomous and governed by an executive board and a representative assembly; half the members of each body represent employers, half represent employees. The administrative costs and injury compensation are financed by payments from the member companies. Within each BG, payments vary among companies according to an elaborate system of "hazard classes," under which companies with more hazardous operations pay higher levies.

Every BG has its own "technical inspectorate," experts in

the health and safety problems of particular industries, and issues its own guidelines for industrial operations. BG Chemie, which covers the chemical industry, requires each inspector to have a university degree in engineering and at least five years of experience with a chemical company. To become a fully qualified inspector, a candidate must also complete a two-year apprenticeship and a set of examinations supervised by BG Chemie.

The BGs develop their guidelines in cooperation with member companies and their workers. These guidelines, applicable to all member companies, must be approved by the federal minister of labor and social affairs. Compliance is assured through plant visits by BG inspectors, who act as technical consultants to both management and workers. Like British factory inspectors, their primary role is to be advisers and not police officers, but the inspectors do have the right to see any operations they wish, the right to inspect any documents they consider relevant, the power to issue fines, and the power to halt unsafe operations.

The inspectors are required by law to work closely with the work council and with safety committees at each company. German law requires firms with over twenty employees to have work councils. They are elected by direct, secret vote of the employees and have the legal right to intervene on matters like work rules, starting and stopping times, recruitment, determination of piece rates, transfers and dismissals, and changes in work methods. On health and safety issues, a work council must approve all company health and safety measures, including the appointment of plant physicians and company safety officers. It also has broad access to company information. In fact, the council and the employer are expected to have access to the same body of information concerning health and safety. The BG inspectors are required to assist the work council in exercising its right to codetermination. This means, in particular, that inspectors must inform the council

of any deficiencies they find and describe ways to correct
them.

Both the BG inspectors and the work council are expected
to cooperate with company health and safety committees.
These are composed of workers, management representatives,
and company safety officers. They must be consulted on
health and safety matters, but, unlike the work council, the
health and safety committees generally focus on providing
information and overseeing the company's routine health and
safety activities. BG inspectors are held strictly accountable for
company information they gather in the course of these activities. Some data are kept under tight security, and large fines
may be levied against BG employees who make unauthorized
disclosures.

At the time of the B.F. Goodrich announcement, BG
Chemie had already become active in the VC problem because of its growing concern about the incidence of acroosteolysis and Raynaud's syndrome. In 1972 a member of its
technical inspectorate started a two-year study of VC-PVC
production in West Germany with special emphasis on the
extent of worker exposure to VC and on possible control measures. The technical inspector responsible for the study, Dr.
Werner Ernst, had previously worked for Wacker Chemi-
tronic, a producer of semiconducting materials, which was a
subsidiary of Wacker Chemie, the large VC-PVC producer.
Information gathered for the study included data on production technology and methods the producing companies regarded as proprietary. Therefore, initial access to the data was
limited to a small working group made up of the few BG
Chemie officials directly involved with the VC-PVC study.
Later, the information was shared more widely, as government
officials and independent scientists became involved in the
development of regulations for VC. The study, however, was
never published in its entirety.

The BGs are one half of the West German dual system. The other half falls under the Ministry of Labor and Social Affairs. The ministry issues regulations under West German federal statutes, and these, like their British or American counterparts, have the force of law. Government regulations are generally coordinated with BG guidelines. In most cases, regulations or guidelines cover a particular problem; if they overlap, the stricter requirements take precedence. Occasionally, as in the VC case, guidelines and regulations are complementary, with the BG guidelines spelling out in precise technical detail the requirements of the ministry's regulations.

Regulations are enforced by the ministry's factory inspectors. Like the BG inspectors, they have access to all documents or operations, they must work with the work council and company health and safety committees, they can and do act as advisers to company officials, and they have power to halt or alter operations they consider unsafe. Generally, the BG inspectors and the Ministry of Labor inspectors try to coordinate their activities.

When B.F. Goodrich made its announcement, the Ministry of Labor, like BG Chemie, was already actively involved with the question of workplace exposure to VC. This was because of media attention and parliamentary questions concerning a study of VC exposure conducted at the University of Bonn. The study indicated that VC exposure could damage the lungs, liver, spleen, and vascular system, and researchers called this syndrome the "VC Disease." *Der Spiegel* reported the findings in December 1973, and other publications picked up the story. Headlines included "Alarm in the Chemical Industry" and "Plastics Cause a Grave Disease." On 16 January 1974, six days before the B.F. Goodrich announcement, members of the Bundestag asked the minister of labor how his department would respond. Later in the month, the ministry announced its plan of action. But by this time the B.F. Goodrich announcement had generated further headlines and had demonstrated that the VC problem was even more serious than it had appeared just a few weeks earlier.

VC and PVC companies immediately began urgent large-scale efforts to reduce VC exposure. The small internal BG Chemie study group expanded to approximately twenty members. About a third were members of the BG technical staff; the others were officials from producing companies, a worker representative, occupational physicians, safety engineers, a representative of the Association of Plastics Producers, and a representative of the Ministry of Labor. The key members of the group were three or four of the BG Chemie staff, who maintained close contacts with engineers and managers in the VC and PVC plants. It was this working group, with some changes in membership, that developed the BG guidelines on VC.

In April 1974 BG Chemie published recommendations on VC and informed the producing companies that they should limit VC exposures to 50 ppm. BG Chemie also told companies to plan on reducing exposures even further in the future. A month later, the Ministry of Labor issued a regulation requiring companies to comply with the interim ceiling of 50 ppm that BG Chemie had recommended.

In the Ministry of Labor, responsibility for developing VC regulations fell to the Committee for Hazardous Working Substances (*Ausschuss fur gefahrliche Arbeitsstoffe* or AgA). AgA recommended standards to the ministry for carcinogens used in the workplace. Once a standard was established, companies were legally obligated to comply with it; the standard became part of the government's health and safety regulations, and both the federal inspectors and the BG technical inspectors were responsible for its enforcement. AgA itself was a multipartite advisory committee to the Ministry of Labor, its members specified by law. In 1974 these were:

- Three representatives of producers of hazardous working materials.
- One representative of a company that transports hazardous working materials.
- One representative of the Association of German Institutes for Standards (an association of private standard-setting organizations).
- Three representatives of companies that process hazardous working materials.
- One representative of the Confederation of German Employers' Association.
- Two representatives of trade unions.
- Two representatives of the Worker Protection Authorities of the Länder, of which at least two must be occupational.
- Three representatives of official agencies responsible for research on worker protection.
- One representative of the German Association of Occupational Physicians.
- One representative of the Association of German Safety Engineers.
- One representative of the Federal Office of Health.

These members made up AgA's general committee, but AgA also had six subcommittees, one of which developed standards. This subcommittee established a special working group to handle the VC problem. The VC working group mirrored AgA's own membership: it included officials of VC and PVC companies, BG Chemie, the Ministry of Labor, and independent scientists, and the unions representing chemical workers.

The industry members of the AgA special working group and the BG Chemie working group were chosen after consultation with the West German Association of Plastics Produc-

ers. The association serves the plastic industry by acting as a
clearinghouse for technical information on matters of common concern and by representing the industry in dealings with
government agencies, other trade associations, and the public.
The association was the official spokesperson for the VC and
PVC producers during the regulation of VC and, as in other
countries, organized special committees of scientists, engineers, and managers from member companies to study particular aspects of the problem. From the start of this effort, the
association had continuous contacts with the authorities, as
described by one association official:

> All forms of communication are possible. It is not
> unusual to ring up the man in charge in the ministry
> and go to his office and see him. This is no problem.
> We can also phone him for answers to questions.
> They also contact us with questions because we can
> give them technical information to answer questions
> they get from the public and in Parliament. But when
> it turns out that you are not honest and are attempting
> to deceive them, you and your company or association
> are very quickly out of the dialogue.

Association officials also attended BG Chemie and AgA
meetings and reported to the association's VC subcommittees.
Memos, letters, and reports were sent to government officials,
and these were supplemented with phone calls and meetings.
In addition, between 1974 and 1977, there were countless
other industry-government contacts outside the formal auspices of the association, between producing companies and
BG Chemie technical inspectors, government labor inspectors, and scientists, engineers, and physicians who were part of
either the AgA or the BG Chemie working groups.

Because companies had little difficulty complying with the
50 ppm interim standard, and since AgA did not begin its

deliberations on VC until the second half of 1974, the association did not have to move at full speed—as U.S. producers did—to develop an industry position on VC. Once organized, the association's VC committees pursued two basic goals. One was to defend PVC as a product—to demonstrate to the public and to government agencies that the economic importance of PVC made it necessary to find safe ways to continue its production, rather than phase it out. The second goal was to develop an industry position on further reductions in VC exposures.

After extended discussions, two possible industry positions emerged. One would have set exposure limits in a way that assured the survival of all the companies in the industry. The other option, which industry adopted, was to advocate quicker and sharper reductions in VC exposures and compel the slower companies to follow along, if they could. This option was chosen partly because of concern that industry resistance might prompt even stricter controls. Most companies also preferred to move more quickly, make the necessary investments and adjustments, and then move on to running their businesses. Finally, companies agreed that industry endorsement of a strong effort, despite the risks to marginal operations, would be far more convincing to the public, to the government, and to trade unions. This would help the companies achieve their principal goal of defending PVC as a product.

The industry position was reflected in the standards developed by AgA. AgA announced its first set of VC standards in March 1975, and a second, stricter set in August 1977. Both were announced by notices published by the Ministry of Labor, and because they were quite short (the first was approximately 300 words long), each was supplemented by a detailed guideline from BG Chemie.

Unlike the British and the Americans, the Germans distinguished old VC plants from new ones. They also distinguished PVC plants from VC plants. Under the first standard, any

PVC plant that started operations after 1 July 1975 had to comply with a 5 ppm annual average and a 15 ppm single-hour mean. The plants already in operation before 1 July 1975 had higher limits: an annual mean of 20 ppm and an hourly ceiling of 60 ppm until 1 July 1976, and an annual mean of 10 ppm and an hourly ceiling of 30 ppm thereafter. The separate provision for older PVC plants was made on the grounds that VC was much harder to control there. German officials adopted an annual mean, rather than the hourly, daily, or weekly means used in the other Western countries, because they thought that a hazard resulting from long-term exposures should be controlled by long-term exposure levels. However, they also established an hourly means to limit short-term exposure. These AgA regulations were accompanied by a set of guidelines published by BG Chemie in July 1975 that described changes in work practices and plant engineering that would reduce VC concentrations below the level specified in the standard.

Roughly two years later, on 1 July 1977, the second set of standards went into effect. This was the result of the continuing review of the "state of technical development" that AgA had said it would conduct when it issued its first set. The new regulation required that all VC and PVC plants achieve at least a 5 ppm exposure level, calculated as an annual mean. However, PVC plants starting operations after 1 July 1977 had to maintain annual levels of 2 ppm. As before, hourly mean exposures could not be more than three times higher than the standard. This new regulation was also accompanied by detailed guidelines from BG Chemie.

Both these standards were written by the special VC working group established under AgA, which held six to eight official meetings for each of the two sets of regulations. There were also scores of less formal meetings among various members of the working group: to review reports from medical researchers and company engineers, to visit VC and PVC plants, to discuss available monitoring methods, and so forth. Once the

VC working group had prepared its recommendations, one or two meetings of the subcommittee of AgA and one full AgA meeting were held to discuss the working group's recommendations.

All the subcommittees and working groups within the AgA hierarchy attempted to reach decisions based on consensus. By 1980 all of the standards AgA recommended had been established in this way. In the event of a stalemate, officials of the Ministry of Labor said they would continue to press AgA and its subgroups for a generally acceptable recommendation. If the stalemate persisted, even after repeated attempts at resolution, then officials of the Ministry of Labor would step in, review the information that AgA had developed, and establish a standard on their own.

Although insoluble deadlocks had not yet occurred, there had been sharp disagreements on particular matters among members of AgA working groups and subcommittees. These disputes were generally handled by having the contending parties present their positions to the working group and then discussing the disputed issue. Sometimes, BG Chemie helped to resolve conflict by providing independent judgments based on its own technical expertise and its close familiarity with plant operations. For example, one of the most difficult issues AgA faced in the VC case was balancing the cost and benefits. The chairperson of the standard-developing subcommittee explained that "this is a very important issue, but it is difficult to describe just how it is treated—these decisions are usually the result of long and hard discussions in the committee meetings." He added that the "general approach" was

> to see what's technically possible at the moment. If 30% to 40% of the companies can, for example, reach 2 ppm while the other 60% cannot, we would choose a value near 2 ppm. Our aim would be to push all the companies to reach this level but not to bankrupt them.

Participants in the AgA meetings emphasized that their formal and informal meetings were not secret nor even fully private. One official said discussions took place in a "climate of privacy." That is, the meetings were not open to the public or to the press, and there were no American-style public hearings. But documents under review, such as medical or technical reports, were circulated both within and outside the working groups and subcommittees as were draft proposals for the standards. Moreover, the members of the working groups frequently discussed their activities and decisions with the groups they represented.

The BG Chemie working group worked in ways quite similar to those of the AgA group. A small group of BG Chemie officials, who had extensive experience doing inspections in the plastic industry and good working relationships with plant managers and engineers, prepared drafts of possible guidelines. The drafts were circulated to other working group members, comments and suggestions were noted, and meetings with the whole group were held to develop the final versions of the guidelines. As with AgA, disagreements arose on certain issues: the trade union representatives, for example, pushed for especially strict limits on the time that reactor cleaners spent inside PVC reactors. Such disagreements were eventually settled by extended discussion, further fact-finding in the plants, and eventual compromise.

Compliance with the standards and guidelines for VC was not difficult for most PV and PVC producers. As in other countries, large-scale engineering efforts were necessary, and new technology—for cleaning reactors, for stripping away residual VC, and so forth—had to be developed and installed. One VC producer, Dynamit Nobel, which had six of the seventeen West German angiosarcoma deaths, closed its PVC operations. The Association of Plastics Producers estimated that companies spent approximately DM 100 million to control workplace exposure to VC. [24]

By the late 1970s, VC exposure levels had fallen substantially below the mandated levels. New plants had little difficulty maintaining levels below 1 ppm, while older plants, when they did have difficulty, operated around 2 ppm, but otherwise maintained an average near 1 ppm.

Ultimately, the West German producers achieved VC levels substantially below the required levels. This was the result of several factors. First of all, the companies preferred to err on the side of safety, and this meant achieving VC exposures below the mandated levels. Attaining very low levels also demonstrated to employees and to the public their commitment to occupational health. At the same time, works councils, the labor inspectorate, and the BG Chemie technical inspectorate all pressed producers to continue to lower VC exposures. And, once lower levels were achieved in one plant, it was difficult for the managers of other plants to argue that they could not do likewise. Finally, the changes in engineering and work practices that were instituted to control VC exposures could not be finely calibrated to achieve just one particular exposure level, such as 5 ppm, and no more. For example, workers and safety personnel were trained to detect and stop VC leaks, and they did so even if VC levels were 1 or 3 or 5 ppm. There were even reports of competition between shifts at some plants to see which could attain lower VC levels.

The fact that neither industry nor trade unions nor outside groups could seek judicial review of the standards or the BG Chemie guidelines also encouraged compliance. Once published officially, companies had to comply with them. Industry officials also emphasized that AgA's emphasis on consensual and incremental decision making, backed up by the independent assessments of the BGs, substantially reduced the chances of regulations that industry would wish to appeal. Moreover, industry officials thought that such appeals, even if they were possible, would have had little chance of success, if the other members of a special working group, who were highly respected experts, had rejected the position of a company or an industry.

France

The French approach to the VC problem was a variation on the themes displayed in the British and West Germany cases. France, like West Germany, relied on a dual system for regulating workplace health and safety, and one-half of the dual system was a quango, quite similar to the West German BGs. Its industry structure was concentrated: in 1974 there were only three VC and PVC producers, Rhône-Poulenc, Société DAUFAC, and Solvay, who organized their VC efforts through a trade group, the Association of Producers of Plastic Materials. As in all the other countries, the B.F. Goodrich announcement marked a watershed in French efforts to control VC. French companies learned almost immediately of the Goodrich findings and took urgent steps in January and February 1974 to bring down VC exposures; both halves of the French dual system became active.

In France, the first part of the dual system is the Ministry of Labor, responsible for the enforcing of a large body of health and safety laws and regulations. French regulations are established by decrees (*décrets*) published by the ministries of the French government, under the authority of statutes (*lois*) that have been adopted by the French National Assembly. Compliance with regulations is binding under penalty of law.

The ministry has its own factory inspectorate, which investigates serious industrial accidents, advises and supervises company health and safety committees, and enforces the ministry's health and safety regulations. Inspectors have the right to see any company record relating to health and safety and can require any technical analysis, measurement, or investigation they consider useful. Anyone interfering with inspectors' activities is subject to prosecution. Although they have strong enforcement powers, the principal role of the inspectors, as in Great Britain and West Germany, is to encourage management and labor to follow appropriate safety practices. For example, when inspectors conduct an inquiry into a serious accident, they aim less at fixing blame than at determining the causes and finding ways to avoid recurrences.

The other half of the French dual system is a quangolike body, the National Sickness Insurance Fund (the Caisse Nationale de l'Assurance Maladie, or CN), which has responsibility for compensating injured workers and for establishing measures to prevent injuries. It is financed by contributions from employers. As in West Germany, a complex contribution scheme matches the amount of a company's contribution to the risks of its industry and to the risks created by the company's particular operations. A joint administrative board runs the CN, with an equal number of members appointed by employers' associations and by labor unions. The Ministry of Health and the Family and the Ministry of the Budget closely supervise the CN, and both also have representatives on its administrative board.

The prevention activities of the CN are the responsibility of fifteen National Technical Committees. Each of these is bipartite, with nine employer and nine employee representatives. Like the BGs in West Germany, each National Technical Committee oversees a particular industry or sector and is responsible for studying all questions relating to prevention, statistics, insurance, and contribution levels. The National Technical Committee of the Chemical Industry was the body that developed the French recommendations for reducing VC exposure.

The CN has its own inspectorate, called a prevention service, consisting of more than 500 consulting engineers and safety controllers. They have full access to company records and operations and are basically technical advisers to management and to the health and safety committees in companies. In particular, they help companies comply with the recommendations that the CN technical committees issue. Officials of the prevention service cannot initiate legal action against violators of health and safety laws or CN recommendations but they can recommend that Ministry of Labor factory inspectors do so. Furthermore, if a member of the prevention service discovers a dangerous situation, and if the employer refuses to take appropriate action, the consulting engineer or safety controller may recommend an increase in the firm's contribution to the CN. The prevention services are administered regionally by bipartite, employer-employee administrative boards.

Both the CN and the Ministry of Labor coordinate their efforts in several ways. By law, both are obligated to work closely with the health and safety committees in French companies. Since 1947 such committees have been required by law in all industrial companies with more than fifty employees. The head of the company or his representative is the chairman of the committee. Other members must include the medical officer, the person responsible for training, and as many as three representatives of management and six labor representatives, depending on the size of the company. Employees choose their representatives through the company work committee. The health and safety committees investigate accidents and cases of occupational disease; they propose health and safety measures to management and monitor compliance with the CN guidelines and Ministry of Labor regulations; they oversee company programs for safety training and for provision of health and safety; and they inform the factory inspectors and the CN prevention service of any incidents relating to their responsibilities.

The Ministry of Labor and CN also coordinate their activities through the National Research and Safety Institute (*l'Institut National de Recherche et de Sécurité* or INRS). It provides the Ministry of Labor, the CN, and company health and safety committees, with research, technical assistance, and training programs. The administrative board of the INRS includes representatives of the major French labor unions, the National Council of French Employers, and the Ministries of Labor, Health, and the Budget. The INRS funding comes from the CN.

All three organizations—the INRS, the CN, and the Ministry of Labor—had major roles in resolving the VC problem in France. The other important actors were the representatives of the VC and PVC workers and an ad hoc VC committee organized under the Association of Producers of Plastic Materials headquartered in Paris, representing approximately thirty-five French plastics producers. The association provided its members with a forum for discussing common technical problems; it gathered and published statistical information relating to the plastics industry in France and abroad and arranged for industry representation before government agencies. In addition, it organized industry efforts to protect and enhance the public image of plastic products. The permanent staff of the association was small—three officials and two secretaries. Most of the association's business was conducted by committees made up of member companies' representatives.

The Association of Producers of Plastic Materials became active in the VC problem shortly after the Goodrich announcement. From the start, it was clear that the cancer-VC link imperiled the economic viability of the PVC industry as well as the public image of all plastics producers. As a result, the VC and PVC producers formed an ad hoc VC committee and three subcommittees: one to evaluate the medical data on the VC health hazard, one to examine analytical techniques for measuring VC concentrations in the workplace, and another to assess technical methods for controlling VC exposures. Managers, engineers, and scientists from the producing companies were represented on all the committees. The producers agreed that there would be full sharing of all company information relating to the missions of the three subcommittees.

The initial efforts of the VC committees paralleled those of producers in West Germany, Great Britain, and the United States. The medical subcommittee began an epidemiological review of employee work histories and medical records, searching for cases linking VC with cancers and specifically with angiosarcoma of the liver. (As in other countries, these records, especially for workers exposed to VC one or two decades earlier, were frequently incomplete or unclear.) Members of the other two committees shared information on ways of quickly "plugging leaks" in VC and PVC production systems and on ways of measuring VC concentrations. Most of the company officials responsible for handling the VC problem had virtual carte blanche for purchasing new equipment and investing in new technology to control VC. Most of the VC committee members also had extensive experience with VC or PVC production. Finally, as in the United States, Great Britain, and West Germany, workers at each producing company were kept informed of developing information about the health hazards of VC and about company plans to reduce exposure.

During 1974, as industry was organizing its VC efforts, the INRS took the first steps in creating French regulations for VC. In mid-1975, it published a three-page bibliographical review of the literature on VC as a carcinogen. The authors were a physician and an engineer associated with the INRS. They stated that exposure to VC should be controlled as rigorously as possible and that VC levels should be kept below 25 ppm, and preferably close to 10 ppm. Later in the year, the INRS published a second article that described the VC hazard in detail and gave a general account of preventive measures. The article recommended that companies maintain, provisionally, a 5 ppm daily average exposure and a 15 ppm ceiling.

Neither of the limits recommended in the INRS articles was legally mandatory.

CN entered the picture in early 1976. The National Technical Committee for the Chemical Industry issued a brief technical notice on VC exposure in March 1976. It established two sets of limits for VC: plants built after 1 January 1976 had to maintain average VC levels of 1 ppm and a ceiling of 5 ppm; plants built before 1 January 1976 had to maintain an average of 5 ppm and a ceiling of 15 ppm. Less than a year later, in February 1977, the committee published a formal recommendation on VC exposure. It suggested the same exposure levels as in the 1976 technical notice but added that these levels were only temporary and were based solely on practical considerations. The recommendation stressed that VC concentrations should be steadily and vigorously reduced toward zero because it was impossible to establish a safe exposure level for a carcinogen. It also described monitoring devices for VC. The technical notice and the official recommendation, like other CN guidelines, did not have the force of law. However, companies that did not meet the levels recommended by the CN could have their CN contribution raised. Furthermore, if a worker became ill and his employer had not complied with the recommendations, this fact could be used against the employer in any legal proceedings. Finally, company health and safety committees, as well as labor unions, exerted pressure for compliance with CN standards.

The initial set of French VC standards—the technical notice and the recommendation by the CN—were prepared by a special working group, formed under the auspices of the national Technical Committee for the chemical industry. The committee itself had eighteen members, nine chosen by the labor unions that represent chemical workers, and nine chosen by the National Council of French Employers, after consultation with industry associations for the chemical industry. Since the committee did not have technical expertise on special aspects of chemical industry operations, it set up a special working group to draft the technical notice and recommendation for VC. The most active members of the group were Jean-Claude Thomas of Rhône-Poulenc, the largest French producer of VC and PVC, Michel Odet, a career official of the *Confédération Générale du Travail* (CGT), and Alex Kreyenbuhl, the General Secretary of the National Technical Committee for the Chemical Industry and a chemist by training. Before joining the CN, he had taught organic chemistry at a university and had done research on plastics for both companies and the government.

The group met for the first time in June 1974. From the beginning, a few issues dominated the group's discussions. One was the question of how far VC concentrations could be reduced in existing plants. Industry's initial efforts to "plug leaks" was already achieving exposure levels well below 50 ppm, but there was no way of knowing what further reductions could be achieved. On the other hand, there was little question that new plants could achieve VC exposure levels in the very low single digits, so setting levels for them was not a difficult issue. The initial approach to the issue of VC levels for existing plants was basically "try and see": companies would continue their efforts to reduce exposures—through changes in work practices, improved maintenance, and development of new technology—and the working group would assess the progress. Reports on industry progress were provided by Thomas and by the consulting engineers of the regional CN offices who visited the VC and PVC facilities.

The second vexing issue was establishing a satisfactory working relationship within the group. The first meeting started on a rough note: CGT union representatives described the VC issue as a classic confrontation between capitalists and workers. (The CGT is a communist union, officially committed to the overthrow of capitalism.) They demanded the 1 ppm exposure limit that OSHA had proposed. In the view of Jean-

Claude Thomas, relationships gradually improved after the first few meetings as the different parties realized that they did share the objective of very substantially reducing VC exposures, even it they could not agree on what levels could or should ultimately be achieved. Furthermore, Thomas emphasized that his credibility as an industry spokesperson was strengthened by personal relations that the working group meetings made possible. In his words:

> In large measure, it was a problem of individuals, a problem of developing personal credibility. I eventually got to the point where I had about 80 percent credibility. If I said that something could or could not be done, I would probably be believed. And it was important that what I said could often be confirmed by the CN engineers who visited the plants. They could see what industry was doing, and they could make judgments about what was possible. There were independent ways of confirming what I was saying. As a result of this, the industry experts at the meetings of the adjunct groups were generally treated as experts. They could be trusted to put the facts on the table, and then we could have a very good exchange of views.

The meetings themselves were not open to the press or to the public, but it was assumed that all parties would report back to the groups they represented on the discussions and decisions. The satisfactory personal working relationships did not eliminate accusatory public statements. A CGT publication called *Syndicalisme* charged that PVC workers were continuing to be treated like "common guinea pigs," despite the working party efforts. Furthermore, several months after the working group meetings had started, the CGT sent a letter to the Ministry of Labor charging industry with attempting to corrupt the process of developing new VC standards and demanding that the government impose a standard as strict as the

American one. Thomas regarded the letter as an exception to a generally cooperative process and believed it was simply "a little public flag-waving" that did not disrupt the progress of the working party. According to Mr. Kreyenbuhl, the working group meetings generally proceeded "very reasonably and with few severe disagreements," since all the parties recognized the difficulties of reducing VC concentrations to very low levels and were acquainted with the industry effort that was under way.

Ultimately, the exposure levels proposed by the working group were the results of compromises between the preferences of industry and labor, tempered by reports on VC levels already achieved in plants and on probable reductions in the near future. The draft of the CN recommendations, and of the technical notice that preceded it, were reviewed by regional CN officials and by the full National and Technical Committee for the Chemical Industry. These two documents formed the basis for the decree on VC issued three years later by the Ministry of Labor.

Two features of the way in which the Ministry of Labor developed the final French VC regulations merit attention. The first is that the ministry did not publish its VC decree until the spring of 1980. This was by far the latest official regulatory act for VC in any of the five countries under study. The second was the reliance of the ministry on a second multipartite body, the High Council for the Prevention of Occupational Risks (the *Conseil Superieur de la Prévention des Risques Professionnels*, or the CSRP), for evaluation and approval of its VC decree.

The Ministry of Labor delayed publication of this decree for several reasons. These were described by the general secretary of the CSRP:

First of all, we knew that the VC and PVC companies were already making a great effort to control

VC levels. We knew this because Ministry of Labor inspectors had visited the plants and because the CN was actively involved in the problem. There was no urgent reason for us to take action. Second, once a decree is enacted, it is strictly enforced. Violators can be fined heavily or sent to jail. We wanted to know that the companies would be able to comply with the strict requirements of the decree. We also did not want to publish the decree and then learn that parts of it were inappropriate or did not make technical sense. By waiting, we reduced the possibility of any of these things happening. We wanted to wait until the facts were fully available.

Third, we wanted to wait until the EEC produced its VC directive, which it did in the summer of 1978, so that we could prepare a decree that was compatible with the EEC directive. As an EEC member, we must harmonize our national laws with its legislation.

The first step in developing the VC decree was taken by ministry officials who prepared a draft based on the CN recommendation, the new EEC directive, and the results of company efforts to control VC in older plants. By this time, in 1978 and 1979, VC levels in older facilities were generally below 2 ppm, so a 3 ppm limited replaced the 5 ppm limit of the CN recommendation. A draft decree with these new limits was then sent for review to all members of the CSRP.

The CSRP is a consultative body to the Ministry of Labor, intended to involve both employers and employees directly in the preparation of regulations and of Ministry of Labor policies on occupational health and safety. In 1979 the CSRP had forty-eight members. Thirteen represented various government agencies as well as the CN and the INRS; ten represented the three major French trade unions, ten represented employees; and fifteen were scientists, physicians, and other officials involved in occupational health and safety, drawn mainly from French universities.

The draft VC decree was reviewed, article by article, by a special committee of the CSRP at two meetings, in September and November 1979. The three men who had been most active in the CN working group—Thomas, Odet, and Kreyenbuhl—attended both meetings. The participants raised many of the same issues that the CN working group had already treated at length: Odet, for example, opposed setting different exposure limits for old and new plants, and Thomas argued that it was technologically impossible for older plants to achieve levels as low as those of new plants. In the end, the discussions produced only a few minor changes in the draft. In principle, the CSRP could have proposed major changes, and then the Ministry of Labor would have had to decide whether to incorporate them in the final decree. If, however, the parties immediately affected by the problem had worked out a mutually satisfactory agreement, then the CSRP and ministry were unlikely to modify it.

The final and official VC regulations were issued on 12 March 1980, when the Minister of Labor published a decree on VC in the *Journal Officiel* (which is the rough equivalent of the U.S. *Federal Register*). Like the CN recommendation, the decree distinguished between old and new plants. Plants starting operation after the effective date of the decree (after September 1980) had to maintain a weekly average of 1 ppm and a ceiling of 5 ppm. Older plants were required to maintain an average of 3 ppm. Measurement and control systems could be installed only after consultation with company health and safety committees, and Ministry of Labor inspectors could require modifications of these systems. The records of actual exposure levels had to be available to the workers, government inspectors, and representatives of the CN.

By the time the French decree on VC became effective, in September 1980, the French VC and PVC industries had already been in compliance with the decree for several years. The total cost of compliance was approximately 80 million francs. At no point in the process that led to the VC regulations was there a formal assessment of costs and benefits. In the end, according to Kreyenbuhl of the CN, a limit of 3 ppm for older plants was chosen because a lower limit—such as 1 ppm—might have forced older facilities to shut down. An official of the Ministry of Labor said the ministry's general approach to the costs and benefits of control in carcinogens was to "try to make some progress, see what the costs and results have been, and then generally push for further progress."

Since industry had been in compliance with the VC regulations long before they became effective, companies had no reason to seek judicial review of the regulations. In principle, any person or group whose interests are adversely affected by a regulation of a French government agency can seek a review and possible annulment of the regulation in the French administrative courts. These courts handle all litigation to which a public authority is a party. However, litigation rarely followed government decrees on workplace health and safety for the chemical industry since the regulations were established generally only after industry had complied with them.

Chapter 5
JAPAN: COOPERATION WITHOUT QUANGOS

The Japanese approach to the VC problem was, in effect, a hybrid of the institutional arrangements and operating policies displayed in the United States and European cases. The Japanese VC-PVC industry, like the American, comprises a large number of competitive firms. Before the VC episode, they had little experience cooperating with each other or with government on workplace health and safety issues. In fact, one member of the Japan PVC Association—the Chisso Company—was responsible for the pollution that led to the Minamata episode, perhaps the most bitter environmental conflict in Japanese history. But like the three European countries, the Japanese relied on both regulations and guidelines to control

VC; government inspectors who acted as advisers, and not like police officers, were responsible for overseeing compliance with the regulations and recommendations; and the Japanese approach to VC emerged from myriad small group meetings and informal exchanges of information.

Above all, the most striking feature of the Japanese approach was that business and government cooperated on the VC issue without any mediating institution such as a quango. Their relationship was direct, as in the United States, and not a business-quango-government relationship. In Japan, the Ministry of Labor supervised the development of regulations and guidelines for VC because the ministry, like OSHA, did not share its responsibility for workplace health and safety through any sort of dual system. The ministry's role resembled the Department of Labor's, in part, because the U.S. Occupation established the ministry in 1947 and patterned it on U.S. institutions. (Before the war, a branch of the Ministry of Health and Welfare had been responsible for labor matters.) The chief difference between the ministry's role and OSHA's was that, under Japan's parliamentary system of government, the minister of labor is accountable to the majority party in the Japanese parliament. One of the ministry's basic responsibilities is to prevent workplace health and safety problems by issuing its own regulations and guidelines and by encouraging employers and workers to undertake voluntary efforts. The ministry's authority derives from the *Labor Standards Law of 1947* and the *Industrial Health and Safety Law of 1972*. These empower the ministry to issue ordinances, such as the "Ordinance on Prevention of Injury from Specific Chemicals," covering workplace health hazards. The 1972 law, in particular, encouraged preventive measures. It clarified the responsibility for workplace health and safety within companies and encouraged companies to develop and follow voluntary health and safety measures, rather than focus solely on complying with ordinances.

To translate these two broad statutes into practice on a particular problem such as VC, the ministry uses a combination
of regulatory instruments. One is formal regulation of a
hazardous substance under the provisions of the relevant ordinance. Such regulations generally take the form of brief statements of technical requirements. They are the rough equivalent of the U.S. and German regulations, the French decrees,
and approved codes of practice in Great Britain. For workplace
hazards, they specify exposure limits enforceable under penalty of law. The other regulatory instruments are technical
guidelines (also called circular notices) that recommend ways
to limit exposure to hazardous substances. Such guidelines are
based on extensive consultation with industry and summarize
the judgment of industry and government experts on the most
appropriate procedures for dealing with a particular problem.
Companies deviating from guidelines are not subject to legal
penalty.

The Japanese resolved the VC problem under the guidance
of two technical notices, one issued in 1974, the other in
1975. It was only after the problem had been essentially resolved—that is, after workplace exposures had dropped to the
very low single digits—that the Ministry of Labor issued formal regulations for VC. In this respect, the Japanese followed
the European pattern and relied chiefly on relatively informal
instruments of administrative guidance, and not as the United
States did, on formal regulation.

The ministry assures compliance with its regulations and
guidelines through its 3,000 factory inspectors. Their principal
role, like that of the European inspectorates, is to provide
guidance and assistance to company officials and to company
health and safety committees, and to supervise their activities.
A company with more than fifty employees must establish a
health and safety committee. Half the committee members
must be chosen by the unions, if the company has unions, and
half by management. The committees discuss health and

safety issues, inform workers about company efforts, make recommendations to management, and oversee the company's compliance with laws, regulations, and collective bargaining agreements on workplace health and safety. Worker members of these committees receive technical information from their companies and from the labor union confederations to which their union belongs. The law also requires that companies appoint a general health and safety supervisor (and, in large plants, one or more subordinate health and safety supervisors) who has responsibility for overall control of health and safety efforts within the company's plants. These supervisors are members of the workplace health and safety committee, and they must be approved by the ministry, which can also dismiss them or require that a company increase their number. The law also requires companies to notify the ministry within thirty days of any change of machinery, operations, or work practices that could affect the health of workers. The ministry can then modify or suspend changes it considers harmful. Ultimate responsibility for the health and safety of workers is in the hands of employers, even for injuries caused by the habits or direct actions of workers.

The ministry's responsibilities are not enforced solely through advice and guidance. Japanese health and safety inspectors can enter workplaces at their discretion to ask any questions they think appropriate, inspect records, documents, and any other articles, take measures of the working environment, and collect, without providing compensation, samples of products, raw materials, and tools. An inspector who finds a serious problem can issue a warning to the company, and the company must comply with the terms of the warning before the inspector's next visit. If there is a violation of "an especially vicious nature," the inspector may turn the case over for prosecution. Managers and workers may be punished for not complying with the warnings, orders, or regulations. The penalties include up to three years in prison and fines of up to 500,000 yen. An inspector who finds an immediate threat to worker health or safety can order that operations be halted.

The first steps taken by the ministry in connection with VC relied on guidance rather than regulation. Findings reported at the 1969 International Congress on Occupational Health, held in Tokyo, had linked VC exposure to acroosteolysis. In response, the ministry asked the Japanese PVC Association to sponsor a survey of worker health in VC and PVC companies. The result was two separate examinations of all 1,900 workers in the industry, conducted late in 1969 and early in 1970. The survey uncovered only two suspected cases of acroosteolysis. Nevertheless, the PVC Association prepared, under the ministry's supervision, a "Guide to Vinyl Chloride Operation and Use." It recommended procedures for degassing polymerization vessels and suggested that companies install water jets to clean reactor walls and reduce the amount of PVC residue that had to be scraped off by hand. The guide strongly recommended annual physicals for all VC and PVC workers. All the Japanese VC and PVC companies complied with these recommendations. By the end of 1970, every company had installed water-jet cleaning systems in their polymerization vessels. As a result of their efforts, workplace concentrations of VC fell to between 50 and 100 ppm during the early 1970s.[25]

The Japanese took no further steps to control VC exposure until the B.F. Goodrich announcement. Reports of the announcement reached Japan on two fronts. The major chemical companies received press releases summarizing B.F. Goodrich's findings, and they immediately passed the information to the PVC Association, and within days of the announcement, the Japanese press published accounts of the VC-cancer link. These reports triggered sharp responses from government, industry, and labor. Within two weeks, the ministry and the Japanese VC and PVC producers had taken action—and in both cases the first steps involved the Japan PVC Association. During the first week in February, representatives of all producing companies held a discussion of the problem at the association's headquarters. And one week later, the ministry officially requested the association's cooperation in resolving the problem.

The Japan PVC Association represented twenty-two Japanese VC and PVC companies. Based in Tokyo, it had about ten permanent staff members and was financed by contributions from member companies in proportion to their VC and PVC production capacity. The names of its committees suggested the association's major activities: the Technical Committee, Environmental Committee, the Raw Material Committee, the Demand Research Committee, and the Overseas Research Committee. Committee members were officials of member companies. The executive director of the association was usually a former official of the Ministry of International Trade and Industry (MITI). His role was to provide a strong link between association members and the sections of MITI that supervised the chemical industry.

By April the association had organized an Industrial Hygiene Committee to deal with the VC problem. This committee had two subcommittees, one for medical issues, the other for engineering matters. These two subcommittees, in turn, established special task groups to study specific technical problems such as techniques for detecting VC. The members of the task groups and subcommittees were drawn from the senior technical and medical staff of member companies. The association had three overall objectives: to resolve the VC problem "within the context of further development of the VC/PVC industries," to "exchange frankly all information among member companies," to conduct discussions with government authorities and labor unions to "obtain mutual understanding."[26]

To meet its second and third objectives, the Industrial Hygiene Committee relied on extensive communications with officials from labor unions and the Ministry of Labor. During the two years in which the VC problem was resolved, the association was the center point of an informal network of discussions involving all the parties to VC problems. The parties communicated by telephone conversations, informal meetings, and through exchanges of memos, studies, and documents. During 1974 most of these discussions were bilateral: officials of the PVC Association dealt directly with labor union representatives or with officials of the Ministry of Labor. Discussions with trade unions took place at three levels: with members of the company health and safety committees, with officials of the enterprise unions at member companies, and with officials of the ICEF-JAF, a national organization of Japanese labor unions. The government officials who participated in the informal VC discussions were members of the Labor Standards Bureau of the ministry and its departments, such as the Labor Sanitation Department. Formal, multipartite meetings among the parties did not begin until December 1974.

Informal, bilateral exchanges with labor and government were not unusual for the PVC Association and its members. What was unusual, according to several company officials, was the degree of trust and candor in the discussions between the companies and officials of the ministry. Before the VC episode, they said, relations with the Ministry of Labor had been colored at times by mistrust, and some company officials had considered it safer to provide the ministry with minimal information in hopes of limiting the ministry's ability to act.

The VC problem was handled in a different spirit, for three basic reasons. One was the unambiguous gravity of the problem: workers, who were members of the "company family," were threatened by a deadly disease. (For several years after the Goodrich announcement, the Japanese press reported the mounting toll of VC-related deaths in other countries.) Second, member companies knew the steps the ministry might take if the industry did not take a strong lead in resolving the problem. The ministry could have moved swiftly and almost unilaterally to impose very strict standards for VC, stricter, in fact, than the 1 ppm standard that OSHA had proposed. And even if a strict standard took the form of a technical guideline, it would nonetheless, have the force of law—even though violations could not, in principle, lead to prosecution—and companies could not seek judicial review.

The third factor was that nearly two-thirds of the Japanese VC and PVC companies had been involved at least indirectly in the fifteen years of controversy, publicity, and recriminations caused by the outbreak of Minamata disease during the 1950s and 1960s. Victims of Minamata disease, a form of mercury poisoning that causes degeneration of the nervous system, may suffer tingling and numbness in their limbs, impaired ability to move, slowed speech, deterioration of vision, brain damage, as well as genetic and fetal damage; ultimately, many die. The first cases appeared among fishermen and other residents of the village of Minamata. Investigations linked the disease to organic mercury, a waste material from acetaldehyde production, which was dumped into Minamata Bay by the Chisso Company, a member of the Japan PVC Association. Many other Japanese VC and PVC companies either produced acetaldehyde—a plasticizer used to make plastics, drugs, and perfumes—or else purchased it from another company.

The Minamata episode was an environmental watershed for Japanese society. It involved widely publicized accusations that Chisso had concealed early studies linking the effluent from acetaldehyde production to mercury poisoning. In 1969 Chisso was brought to trial by victims of Minamata disease seeking compensation. The four-year trial concluded with the finding that Chisso had indeed released contaminating effluent into Minamata Bay until 1968, when the production method requiring mercury became obsolete. The company had installed a removal system for mercury several years earlier, but evidence indicated that the system was ineffective, and that company officials knew this. The court ordered Chisso to pay victims compensation that ultimately totaled roughly $100 million and nearly bankrupted the company.

The Minamata episode attracted extensive publicity, as photographs of grotesquely deformed victims were displayed around the world. In Japan, a major consequence was enactment of the Basic Law for Environmental Pollution Control in 1970. The law included standards for the industrial use of mercury that were so strict that many companies stopped using it entirely.

As a result of these factors—the cancer threat, the possibility of swift, unilateral government action, and the Minamata episode—the special committees of the PVC Association moved quickly to develop preliminary guidelines for VC exposure, and from the very beginning, the committees maintained close contacts with labor unions and Ministry of Labor officials. By the end of May 1974, the Industrial Hygiene Committee of the PVC Association had developed temporary emergency standards for VC. In a letter to its members, the association urged that companies take these steps:

1. Maintain VC concentrations below 50 ppm in all workplaces.
2. Require workers to use protective devices when VC exposure levels exceeded 50 ppm.
3. Conduct intensive medical surveillance of all employees exposed to VC or PVC manufacture. The examinations were to be conducted by plant physicians following the guidance of the medical subcommittee of the Industrial Hygiene Committee.

The letter to member companies also described the proposed OSHA standard of a no-detectable level for VC. The association also said that it would continue to review medical and technical findings, along with proposed regulations, from overseas and would take measures in the coming months to encourage further reductions in VC exposure levels.

On 24 June the Ministry of Labor issued emergency technical guidelines for VC. These were essentially identical to the guidelines that the PVC Association had recommended one month before and encouraged manufacturers to reduce VC exposure as far below 50 ppm as possible. The similarity was the result of extensive discussions between the ministry and the PVC Association.

In the following months, VC and PVC companies continued their efforts to lower VC exposures. As in the other countries, this involved sealing leaks in the pipes and PVC reactors and limiting the time workers spent in areas where VC was produced or handled. After reviewing the progress companies made during the summer, and after discussions with the ministry, the PVC Association issued a second set of voluntary guidelines in September 1974. These urged companies to reduce VC exposures to 25 ppm as soon as possible.

In November, after further consultation with the ministry and labor unions, the association announced a third set of guidelines recommending that by April 1976 companies reduce VC levels to a time-weighted average somewhere below 10 ppm and also maintain, at all times, a 10 ppm ceiling on VC levels. Like the first two sets of guidelines, these were announced only after discussions indicated that companies would achieve these levels by the April 1976 target.

In developing their three sets of guidelines, the subcommittees of the association reviewed information about the efforts in other countries to control VC exposure. Some information came directly from chemical companies in the United States and Europe, some from government or from union officials who attended international conferences on the VC health hazard, and an ongoing stream of general, VC-related information came from frequent reports in Japanese newspapers on political and medical developments in other countries. The PVC Association also gathered first-hand information about the efforts overseas by sponsoring two missions, one to the United States, and one to Europe, to interview individuals working on the VC problem.

The missions had two major consequences. One was a widely shared conclusion that the Japanese companies should act as if there were an unquestionable causal link between exposure to VC and cancer, even though further studies were needed to determine the precise nature of the link and the exposure levels likely to cause cancer. The other consequence was strong collaboration among occupational physicians and engineers from different companies. Before the VC episode, such cooperation had been common only within a particular company; collaboration across companies had been difficult and infrequent.

While it was expanding its contacts with companies and governments overseas, the PVC Association also held discussions with other Japanese trade associations. From the start of the VC problem, the Basic Industry Bureau of MITI encouraged the association to coordinate its activities with the trade associations of companies that used PVC as a raw material, such as the Nippon Vinyl Industry Association and the PVC Pipe and Fixture Industry Association. MITI's concern was to reduce the economic disruptions that the VC problem threatened to create for downstream companies and for their consumers. A wide variety of industries depended on PVC in Japan, and they employed approximately 500,000 workers. MITI focused its attention on the PVC Association for a single reason: if its members could remove VC from the PVC that they sold, then PVC buyers would not have to take costly steps to reduce worker and consumer exposure to VC. As a result, discussions among officials of the PVC Association, the downstream trade associations, and MITI continued until late in 1975, when it became clear that PVC producers could "strip"

VC monomer from the PVC that they sold. This breakthrough resulted from an effort by a joint research team established by the PVC Association and financed by MITI. Once
the problem of residual VC was solved, the PVC Association
published a public relations pamphlet entitled "PVC Products
Are Safe."

While these industry efforts were under way during late
1974, officials of the Ministry of Labor held internal discussions of possible revisions of the ordinances regulating hazardous chemicals and of possible technical guidelines for VC.
The ministry was also gathering its own information on the
VC problem and on control methods. Some information
came from reports prepared by government agencies overseas;
some was obtained when ministry officials attended international conferences. Other information came through discussions with Japanese VC and PVC companies, and from the
government's inspections of plants and medical examinations
of workers. Both the inspections and the medical examinations
were conducted by the prefectural offices of the ministry. Later
in 1975, as the ministry moved closer to producing a second
set of technical guidelines, it circulated draft guidelines and
proposed revisions of ordinances to companies and trade
unions and discussed the drafts with their representatives.

The Ministry of Labor also held eleven bilateral meetings
with labor union officials between August 1974 and June
1975. These generally involved two or three top officials from
the Labor Standards Bureau or the Industrial Health Division
of the ministry and representatives of the Japanese Chemical
Workers Union and the major enterprise unions at VC and
PVC companies. The meetings were private discussions held
at the ministry offices in Tokyo. The agendas covered reports
from other countries, regulations—such as OSHA's 1 ppm
standard—adopted overseas, the results of medical examinations of Japanese workers, technologies for reducing VC, and
specific technical issues (for example, should workers be re-

quired to wear face masks), the number and placement of measuring points, and timing of further VC guidelines. In these discussions the union representatives pressed for VC standards comparable to those in other countries.

At one meeting in late 1974, the union representatives suggested that the ministry conduct formal, tripartite discussions among labor unions, the ministry, and representatives of the PVC Association. The PVC Association agreed to this proposal since it knew the ministry would eventually issue further technical guidelines, and these could be discussed at the proposed sessions.

The first tripartite meeting was held in December 1974 at the Industrial Safety Laboratory of the Ministry of Labor. Five officials of the ministry, thirteen representatives of labor unions (including representatives from many of the major enterprise unions), and eight representatives of the PVC Association attended. A government official was moderator. The opening comments of the government representatives stressed that all parties agreed with the basic principle of "no more sacrifice" of employees' health. They also said the ministry was working on changes in ordinances and new technical guidelines for VC. The union representatives summarized their position in several short principles. These included "don't make the human being a guinea pig," "there is no choice between life and job, but a right to both health and work," and "develop standards from an international viewpoint." The representatives from the PVC Association stressed that the problem involved a serious threat to human life and that there must not, in the end, be any differences of opinion between labor and management. But association officials also noted that factors like company profitability also had to be considered. They described association efforts to gain information about the VC problem and to develop ways of mitigating it. Later in the first meeting, one official from the ministry stressed that any information developed by companies should be open to the public so that industry's efforts to control the problem would have genuine credibility. The official also chastised the industry by saying that at least some companies had shown a tendency toward secrecy in the past, and he said he hoped that this would not happen in this case. The parties also agreed to send a joint industry-labor research mission to Europe. The mission was conducted in February 1975. The participants included one representative from the PVC Association, three from member companies, one from the ICEF-JAF, and four from company unions.

The next full meeting of the tripartite group did not take place until June 1975. Like the other meetings, it was chaired by an official of the Ministry of Labor and was attended by many representatives of the ministry, member companies, and their labor unions. One of the principal topics of discussion was the next set of technical guidelines that the ministry would eventually issue. The parties agreed that these guidelines would only be temporary and could be revised as further information became available.

Between the two formal tripartite discussions, informal meetings of the parties continued. Discussions also took place in several new forums, including several conferences of the Japan Chemical Industry Association and the Chemical Workers Union at which representatives of the PVC Association discussed their efforts to control VC exposure. Others took place between December 1974 and February 1975, when representatives of the Ministry of Labor, VC and PVC company unions, and the PVC Association made joint visits to several VC and PVC plants to monitor a study of VC exposure levels.

In June 1975 the ministry issued a second set of technical guidelines for VC. Like the first, these were written by its staff members. Drafts of the guidelines had been circulated in advance among members of the PVC Association, and among trade union representatives. The new guidelines included several major changes from the first set.[27] These were:

- VC concentration should be limited to a geometrical mean[28] of 2 ppm or less.
- The upper limit of the geometrical standard deviation for the geometric mean[29] of VC exposure should not exceed .4.
- VC exposure should be measured at a minimum of five points in every workplace, measurements should be taken for at least ten minutes and at least once a month.
- Workers should not enter a polymerization vessel unless VC levels were below 5 ppm.

The recommended VC levels were to be achieved "as soon as possible." Unlike the French and German VC standards, there was no distinction between old plants and new plants.

These guidelines were chosen on the grounds that they were consistent with available medical evidence and because it was clear that companies could comply with them technically and financially. One indication of technical feasibility is that measurements of workplace exposure to VC made between December 1974 and February 1975 had shown that roughly 85 percent of the workplaces already had geometrical mean exposures of 2 ppm, and 97 percent had achieved levels below 3 ppm.[30] Officials of the ministry and the PVC Association emphasized that the "spirit" of the guidelines was that the companies try to reduce VC concentrations to the lowest feasible levels, rather than curtail their efforts once they achieved a geometrical mean of 2 ppm and a standard deviation of .4.

Economic feasibility was also one of the factors leading to the choice of this standard. At no time in the discussion were any formal cost/benefit calculations made, but companies did provide the Ministry of Labor with extensive estimates of the costs of reaching various exposure levels and the impact of such costs on company profits. But both government and company officials emphasized that all parties knew they were dealing with a carcinogen and therefore, they said, financial factors did not play a primary role in the final choice of VC exposure levels. Nevertheless, no Japanese plants were closed because of inability to comply with the VC guidelines.

By the time the Ministry of Labor announced its second set of guidelines, most companies had already complied with them. Those that had not were expected to achieve compliance in the near future. One factor that helped companies was the sharing of technical information through the subcommittees of the PVC Association. There were, however, limits to this collaboration. One company, for example, had developed a special process for reducing residual VC in the polymerization vessels several years before the VC-cancer problem arose. It did not share this technique with its rivals for competitive reasons. Officials of the PVC Association encouraged the company to disclose its technology, but the company offered only to license it. Several companies overseas eventually did license the technology, but no Japanese companies did so; instead, they developed their own techniques.

The cost of compliance with the guidelines was approximately 22 billion yen, or roughly $100 million. Half was for equipment costs, and the rest for operating expenses. A small portion of the cost was defrayed by a loan from the Japan Development Bank. About 10 percent of the equipment cost was loaned to the manufacturers at 6 percent, a rate about 3 percent lower than prevailing rates on comparable loans. This was the first time that the Japan Development Bank had made a loan for a major occupational health problem (although the Ministry of Labor had provided limited financing to small companies that had difficulty complying with other occupational health regulations). The subsidized loan was provided only after a long negotiation in which MITI and the Ministry of Labor encouraged the Japan Development Bank to provide the funds. Company officials said that the assistance was appreciated but stressed that it was not a major factor in their decisions, since it reduced the cost of compliance by only a small fraction.

By the end of 1975, according to the PVC Association, most companies were attaining levels of VC exposure under 1 ppm, though some companies still had occasional difficulty maintaining the 2 ppm level. Several factors led to this "over-compliance." One was the difficulty of targeting any extremely low level of exposure, like 2 ppm, and achieving it exactly. Therefore, the companies aimed to reach levels below 2 ppm to err on the side of safety and compliance. Second, the guidelines had specified a standard deviation as well as a mean. When companies attempted to reduce the standard deviation, the mean exposure level fell lower and lower. The third factor was that companies made a genuine effort to comply with the "spirit" of the technical guidelines and reduce VC exposure to a minimum feasible level. This effort was encouraged by the fact that inspectors from the Ministry of Labor and members of company and health and safety committees monitored the results of the company's efforts. The degree of company effort was indicated by the fact that workers were rarely being permitted to enter reactor vessels when VC levels in the vessels exceeded 2 ppm, even though the guideline suggested a 5 ppm limit. Furthermore, even though the guidelines did not restrict the time a worker could spend in a reactor vessel, companies tried to keep this exposure to thirty minutes or less.

No litigation followed the announcement of the technical guidelines or the modifications of health and safety ordinances. Company and union officials said that it was possible, in principle, for a dissatisfied party to take legal action, but they explained that such action was extremely rare, since the process of reaching a consensus on standards, as well as the process of developing a strong base of technical and medical information to support the standards, meant that parties had little reason for legal objection. Furthermore, company officials noted that they had an "everlasting relationship" with government officials, and this raised the risk that judicial action could impair future relationships.

Chapter 6
ADVERSARIAL AND COOPERATIVE DECISION MAKING

The decision-making processes presented in the five case studies fall into a definite pattern, one that clearly parallels the pattern of institutional arrangements described earlier. In all the countries with strong hierarchies linked by elaborate networks, business, government, and labor worked together to resolve the VC problem. In the United States, where the hierarchies were weak and networks were virtually nonexistent, they did not. The next chapter explains how different institutional arrangements tend to load the dice in favor of either adversarial or cooperative decision making. But first it is essential to understand precisely what distinguishes these fundamentally different ways of resolving modern industrial problems.

The answers to four ostensibly simple questions reveal the underlying differences. Taken together, they also provide clear definitions of both "adversarial" and "cooperative" business-government relations, terms widely used but rarely defined.

The first question is: *Who, in practice, makes the important decisions?* In the VC controversy, in all five countries, many crucial decisions revolved around what was called "the numbers game": the question of what exposure levels for VC, measured in parts per million, should be recommended or required. This issue dominated the VC episode for several reasons. Above all, the exposure limit was the point at which the complex and uncertain issues of cost, feasibility, and worker protection intersected. Lower ppm levels were generally more costly to achieve, but they also provided greater protection, unless there was a "no effect" threshold below which lower exposures did not matter. The ppm question was also a political lightning rod because it was such a beguilingly simple performance criterion. Other, much more complex matters, such as control and monitoring technology, were equally important for protecting workers, but a demand for infrared spectrophotometry is a less rousing political battle cry than a demand for a 0 ppm exposure to a carcinogen. A ppm limit was also critical to any successful control system. It provided clear guidance to the various individuals who design equipment to precise specifications, develop new work practices, and inspect for compliance; it also offered aid to company officials who may need to defend themselves in court against charges they have permitted hazardous working conditions and to workers seeking assurance that they are actually protected from a hazard.

Who, then, made the ppm decision in the five countries? In the United States, there was essentially a two-step decision process, an approach that appears again and again when government agencies deal with modern industrial problems. In the first step, a small group of agency officials—in this case, the assistant secretary of labor for OSHA, the director of Standards Development, and staff assigned to the case—wrote the final regulation. They made their final decision on VC in private, and it remained a secret until its formal announcement. These officials relied, of course, on the hearing record, on post-hearing submissions, and on their own expertise, but they wrote the VC regulations by themselves, and none of the other parties who would be directly affected by their decision participated in the final decision.

The second step took place in court. Judicial review of agency decisions is not legally mandated, but opportunities to seek judicial review of agency decisions are generally greater in the United States than in the other four countries. However, as the next chapter will show, adversarial relations impel parties to seek court review, and it has become a normal part of U.S. regulatory decision making. As a result, judges make the final decision on how much, if any, of a regulation like the VC standard will become law. Moreover, the judges need not confine themselves solely to determining whether the regulations were developed in accordance with appropriate procedures. They also make direct judgments on highly technical issues. In the VC case, the appeals court concluded that companies had yet to implement many useful techniques, that the industry could achieve VC levels of 1 ppm, and that the VC and PVC manufacturers needed "more faith in their own technological potentialities." The court made these and other substantive judgments in fulfilling its broad obligation to determine whether OSHA's decision was "supported by substantial evidence in the record taken as a whole." In the two-step, U.S. decision process, the key decision makers were agency officials and judges, acting in private.

The agency officials made important decisions on a provi-
sional basis, subject to the intense scrutiny and final verdicts
of judges.

In contrast, in the other four countries, the critical decision
makers were the parties directly affected by the VC problem.
Their decisions were subject to close government scrutiny, but
through an executive agency, not a judge, and the basic decisions, including the requirements for VC ppm levels emerged
from the working group discussions among affected parties.
Executive agency officials moderated some of these discussions, participated in others, and in the end had formal authority to accept, modify, or reject the working groups' decision. But government officials did not determine the final
recommendations or regulations on their own. Trade union
and industry officials along with technical advisers and government officials reached the fundamental decisions about the
VC problem. In France, in fact, officials of the Ministry of
Labor did not even issue their final formal regulation until
several years after the VC problem had been fully resolved in
the workplace.

The second basic question that distinguishes adversarial
from cooperative decision making is: *Who participates in the
process?* In all five countries, a common thread ran through
the professional background of many participants in the VC
decision. Most of them, quite naturally, had scientific training
in specialties like toxicology, epidemiology, occupational
medicine, and chemical engineering. Other leading participants were executives of VC and PVC companies, many of
whom were scientists or engineers, and trade union officials.
Nearly all of these individuals had expertise or direct experience with some substantive aspect of the VC problem.

What distinguished the adversarial U.S. case was an additional class of prominent participants—the lawyers—whose
expertise was legal procedure, not the scientific, technical, or
administrative problems of VC. Lawyers participated from the

very start of the U.S. decision process for VC. At the first meeting of company representatives at the Society of the Plastics Industry, roughly a third of the participants were lawyers, since nearly every company delegation included at least one. Later, lawyers helped build the industry "case" for the OSHA hearing and for the judicial appeals. They cross-examined witnesses at the hearing; government lawyers helped OSHA draft its proposed and final regulations; and, as judges, lawyers made the final decisions on OSHA's VC regulations.

In the other four countries, there were only two individuals with legal backgrounds who played a direct role in the VC episode. One was the general manager of the West German Association of Plastics Producers, who was an attorney. The other was the legal secretary of the British Transport and General Worker's Union, who incidentally did not even have a degree in law. Neither was involved in the VC episode because of his legal background. The only other lawyers in the cooperative cases were the legal experts—such as those of the Conseil d'Etat or the West German Ministry of Labor—who reviewed their country's final VC standards for compatibility with the laws under which they were promulgated.

Another class of participants further distinguished the adversarial and cooperative decision processes. These were individuals and organizations who acted as *intermediaries*. The best examples are Dr. Ernst of BG Chemie in West Germany and Mr. Kreyenbuhl of the French CN National Technical Committee for the Chemical Industry. These two men, and the quangos they represented, were neither fish nor fowl by American institutional standards. They represented neither labor nor management nor government. Instead, they were simultaneously accountable to both labor and management, since both were equally represented on their organization's boards of directors and on their internal operating committees. They also had the role of intermediaries vis-à-vis government agencies. In the words of one West German industry official,

"The BGs tell the government people sitting behind desks what's going on in the factories."

There were no direct counterparts to the BGs or to the CN technical committees in Japan or Great Britain. However, the role of individuals like Messrs. Ernst and Kreyenbuhl was taken, to a significant extent, by the factory inspectors. In the VC case, they provided trade union officials and the VC working parties with independent corroboration of the extent and pace of industry's effort to control VC exposure.

The U.S. case showed that virtually none of the technical, managerial, or labor union participants acted like intermediaries. In theory, the two consulting firms involved in the VC episode could have served as potential intermediaries by developing data and proposals at the behest of more than one party. In practice, however, Arthur D. Little was hired by the Society of the Plastics Industry, and its findings appeared to many observers to be weighted toward industry. The Foster D. Snell study was prepared for OSHA, and it was not completed until shortly before OSHA promulgated its final standard. Had there been negotiations between OSHA and industry, attorneys could perhaps have served as intermediaries. But there were no negotiations, and there were no intermediaries, only partisans aligned on various sides of the issue.

Another difference among the participants was their status within their own organizations. In the United States, the major participants in the VC episode generally held senior positions in their organizations; in the other countries, many important participants came from the middle ranks. At the OSHA hearings, for example, the president of the Society of the Plastics Industry, the assistant secretary of labor for OSHA, and the director of NIOSH all testified. Later, when OSHA decided on its final VC standards, the key actors were Dr. Daniel Boyd, the director of health standards, and John Stender, the assistant secretary of labor for OSHA. In contrast, the members of the VC working groups in Europe and Japan generally held middle management or technical staff positions in their organizations. There were some exceptions to this pattern: some middle-level officials did testify at the OSHA hearings, and in Europe the working parties included a few top officials, such as the head of the British Health and Safety Executive. However, most working group members were health and safety officials from trade unions, operating managers, or the technical personnel from VC and PVC businesses.

The third basic question about the decision processes is: *How did the participants influence the important decisions?* The answer reveals another set of sharp contrasts between adversarial and cooperative business-government relationships. The first and most important difference was that the decision-making process in the United States gave parties only indirect means of influence. Industry and trade union officials could testify at public hearings, seek media publicity, attempt to influence OSHA through Congress or the Office of Management and Budget, threaten litigation, and argue in court. They could not, however, participate directly in the final authoritative deliberations on the contents of the VC regulation. This was the exclusive prerogative of senior agency officials and judges.

In all the cooperative cases, the parties immediately affected had a range of opportunities, extending over several years, for direct influence on important decisions about VC. As a result of the wide opportunities for direct influence, participants developed personal relationships with each other. They came to know each other as individuals, and not solely as representatives of other interest groups. This aspect of cooperative decision making was stressed repeatedly by participants in the VC working parties. A remark of Jean-Claude Thomas of Rhône-Poulenc captured this element: "It was a problem of individuals, a problem of developing personal credibility." Only a few members of the multipartite working groups had even been acquainted before the VC problem arose. Personal relationships developed during the actual efforts to resolve the VC problem through what one official of the British Health and Safety Executive called the "cut and thrust of small meetings," carried on over several years. Under the adversarial approach, however, the principal personal contacts between representatives of the major parties occurred in public: at OSHA hearings and in court. These individuals occasionally became acquainted and somewhat friendly; in fact, their harsh public rhetoric often belied polite working relationships. By and large, though, personal relationships did not develop in the course of OSHA's regulation of VC.

A further difference was that in the United States the channels of communication, and therefore the opportunities for influencing decisions, were almost exclusively public, while in Europe and Japan the working sessions on VC were held in private. The adversarial approach relied essentially on two sorts of meetings. One was fully public: the OSHA hearings and court arguments; and the other was virtually secret: the final drafting of a standard within the agency, and the deliberation of judges and clerks. In the cooperative cases the communication channels could best be described as semiprivate. The meetings were not open to the public or press, no transcripts were available to the press or public, and no hearings were held. It was understood, however, that members of working parties would report to the groups they represented on the discussions and decisions of the working parties.

The meetings among the major parties differed in another way. In the United States, procedural formalities guided and constrained contacts among the parties. This was especially true for both the court proceedings and the triallike procedures of OSHA hearings, which included testimony and cross-examination under the guidance of an administrative law judge. In contrast, very few rules or procedures constrained the meetings of the European and Japanese working parties. In Great Britain, for example, the availability of reports from the subgroups and from the participants determined the timing and agenda of meetings. A government official held the chair, and issues were discussed until the parties were satisfied. Furthermore, official working party meetings were complemented by a much larger number of even less formal meetings—at the offices of trade associations for the chemical industry, on the plant floor at VC and PVC plants, and in government offices. Because OSHA handled the VC problem under emergency rule making, the agency was bound by law to follow a strict timetable. OSHA had only six months to draft a proposed standard, hold public hearings, digest several thousand pages of testimony, and write a final standard. In the other countries, the government officials treated the VC problem like a matter of high urgency, but none was compelled to follow a legally mandated timetable in developing an official VC standard. The French, for example, waited until 1980 to issue official VC regulations.

The formality of adversarial decision making also manifested itself in heavy reliance on documentary evidence. OSHA's VC hearing record exceeded 4,000 pages, and the testimony of Firestone Plastics was nearly 400 pages long. By submitting long and detailed documents, the parties accomplished several objectives. They laid a foundation of evidence for the litigation that regularly follows promulgation of a regulation. They tried to communicate information and judgments about complicated specialized fields of science, medicine, or engineering—which could not be described with sufficient accuracy in short, simple documents. And some parties may have practiced the strategy of "confusion to the enemy" by seeking to bury government agencies in an avalanche of documents.

Compared with the other four countries, parties in the United States had far fewer opportunities to communicate through discussion or visits to VC and PVC plants. OSHA officials made only a few plant visits, such as an industrial hygiene "walk through" arranged by B.F. Goodrich shortly after the company's January 1974 announcement, and three or four one-day plant visits several months later. Aside from the public hearings, these were the principal opportunities companies had to communicate with government officials on site about the practical, plant-level aspects of the VC problem. Most communication between business and government took place via documents and testimony presented in the hearing room of the Department of Labor in Washington, D.C. For those who were utterly unfamiliar with the problem, one witness provided a small jar of PVC pellets and a gas mask, which were later stored, with the rest of the OSHA hearing record, in the OSHA docket office.

Finally, in Europe and Japan, another opportunity for direct communication was explicit, direct discussions of the costs and benefits of possible VC controls. Such discussions were an established part of the decision process and were based on a relatively informal, judgmental approach, not on complex, quantitative models frequently used by U.S. regulatory agencies, although OSHA, at the time of the VC problem, strongly opposed the use of cost/benefit analysis.

The final question about the two types of decision processes is: *How did the pattern of decisions differ?* The first point of contrast is that OSHA made its final decisions about VC exposure limits much more quickly than the government agencies in the other four countries. OSHA announced its emergency temporary standard of 50 ppm only one month after the B.F. Goodrich announcement. OSHA proposed its "no detectable" standard three months later and announced its final standard of 1 ppm after another three months. In contrast, the British VC working group agreed on a final hygiene standard in October 1975, almost exactly a year after OSHA promulgated its final standard. Only in late 1975 did the Japanese Ministry of

Labor make changes in several ordinances to accommodate VC exposure levels suggested in its technical guidelines. The final West German standard was agreed upon in the spring of 1977 and became effective in July 1977. The final French standard did not become law until late in 1980.

The faster pace of government decision making in the adversarial case was inevitably reflected in industry decision making. Within a few months of the Goodrich announcement, U.S. VC and PVC producers had to seek agreement on how far they could reduce VC levels. European and Japanese producers made early commitments to reaching levels below 50 ppm, but further commitments were not made until many months or even years later.

OSHA also differed from its overseas counterparts in that it made a single, once-and-for-all decision on exposure levels. In the other cases, VC levels were regulated through a series of temporary exposure limits of increasing severity, proposed over several years. Furthermore, OSHA's once-and-for-all decision was legally binding, while elsewhere, the interim standards took the form of quasi-official recommendations or technical guidance; companies, in general, were not legally required to comply with the recommendations and guidance.

These two differences in the pattern of decisions resulted in a third important contrast. Because OSHA promulgated regulations more quickly and because it regulated VC in one large step (from 50 ppm to 1 ppm), its regulations were written into law *before* companies had demonstrated the ability to comply with them. In the cooperative cases, both the quasi-official guidelines and formal regulations became effective only when industry *had already complied* with them. In one case, government decisions preceded and, in effect, forced technological developments; in the other cases, the decisions followed the technology. In general, the final VC regulations in Japan and Europe were much more flexible. Great Britain was the most conspicuous case. Its Final Code of Practice for VC recommended a time-weighted average of 10 ppm but also stated that "whenever practicable exposure should be brought as near as possible to zero concentration." In the other three cooperative cases, the final exposure limits were considerably lower than the British ones (and generally higher than the American ones), yet all included the broad recommendation that companies seek the lowest feasible levels. In contrast, OSHA set a simple standard of 1 ppm. This was the most demanding standard—on paper—of any of the five countries. And it was a precise, clear-cut limit: either a company was in compliance with the 1 ppm standard or it was not.

Overall, analysis of the five cases reveals two very different approaches to resolving complex, controversial issues like the VC problem. In the cooperative cases, the parties involved directly influenced the critical decisions and actually made many of them. They achieved consensus through negotiations and discussions among middle-level officials over a period of several years. There were important intermediaries, and very few lawyers, involved in the process. Representatives of the major parties developed personal relationships in small working groups. Meetings were held in private—informally, and without rigid timetables. The working parties explicitly discussed cost and benefit. In the end, the decisions that emerged from these multipartite discussions were less formal, they followed technology, and they included flexible open-ended recommendations that companies should seek to achieve the lowest feasible levels of VC exposure.

In contrast, the key decisions in the adversarial case were made by agency officials and judges on the basis of virtually secret deliberations. Industry and labor did not cooperate but instead took steps to impede and discredit each other. Industry and government behaved similarly. Lawyers were active from start to finish, and no intermediaries acted as buffers among the parties. The various parties had only an indirect influence on the major decisions, and this was largely confined to public, highly publicized, courtlike hearings. Basic information was communicated through documents, and cost/benefit considerations were not an explicit part of the agency's decision process. In the end, OSHA issued its final regulations and required that industry achieve—under penalty of law—a very strict standard, one which exceeded the industry's technical capability at the time it was promulgated.

Chapter 7
HIERARCHIES, NETWORKS, AND COOPERATION

Why did business and government cooperate on the VC problem in all four countries in which networks linked strong hierarchies? Why did this pattern emerge despite important political and cultural differences among the countries? Why did weak hierarchies and networks lead to business-government conflict in the United States?

Answers to these questions lie in the incentives and opportunities created by the two different patterns of institutional arrangements. Networks and hierarchies are not the sole explanation of cooperation, nor do they guarantee it; but they do encourage collaboration. And a clear understanding of how they do so casts light on the questions of whether and how

such cooperation could be brought to bear on the important
U.S. problems for which it is so widely recommended.

Government Hierarchy and Cooperation

Strong government hierarchies appear in all four cases of
cooperation. The four Ministries of Labor were much more
powerful than OSHA, both in their relations with other
branches and departments of government, and in their relations with industry. The ministries had their own captive
scientific branches; their inspectors had greater powers than
OSHA's, and in France and West Germany, quango inspectorates reinforced the efforts of government inspectors; government could hold deliberations in private without requirements
for press conferences, hearings, or disclosure of discussions
and documents; and executive agencies of government, in all
the cases of cooperation, had long histories of power, legitimacy, and prestige at the center of national affairs. Moreover,
government power allied with labor union power, since government inspectors worked closely with company health and
safety committees, which had extensive rights to information,
consultation, and even decision making on occupational
health and safety. And, above all, the executive agencies of
government derived strength from the narrow limits on judicial review of their decisions and procedures.

In contrast, in the U.S. case, the judiciary was stronger and
the executive agency was consequently weaker. There were
many opportunities for judicial review, myriad grounds for
seeking it, and many parties actually did so. The result was
that the executive agency was weaker in relation to business,
labor, public interest groups, and other parties because they
could use court action, or the threat of it, to influence OSHA.
Wide access to judicial review, in effect, dispersed OSHA's

authority among many groups. Under these conditions, parties involved with an issue such as VC face a peculiar set of incentives. In aggregate, these create what is a judicial sweepstakes, open to many participants: the OSHA Act, for instance, grants standing in court to any party "adversely affected by an OSHA ruling." The costs of entering the sweepstakes are principally legal fees, which can be relatively small, both in comparison to the possible gains and to the treasuries of large companies, trade unions, and national public interest groups. Nearly every participant is assured a small prize. Litigation, even if it is ultimately unsuccessful, delays government action, secures publicity, warns government to tread warily in the future, ties up the resources of one's adversaries, and provides litigants with the satisfaction of fighting the good fight. Large wins are also possible. In 1980, after three years of legal skirmishes, industry took OSHA's benzene standard to the Supreme Court, and the court overturned the standard, leaving companies victorious and the agency unsure of how to balance costs and risks in future regulations.

Because of extensive judicial review, OSHA and most other U.S. government agencies are institutional reifications of the American law school maxim that substance arises in the interstices of procedure. It was the judicial influence on the U.S. decision process that accounted for the omnipresence of attorneys in the VC case. Lawyers are the strategists who "build the case" and advise clients on how their statements and actions will look when they are scrutinized, as they almost inevitably will be, by adversaries and judges in a courtroom. Consequently, cooperation with other parties early in the life of an issue can be risky, since it could be construed in court as tacit admission that one's adversaries have valid claims. The Administrative Procedures Act of 1946 set the basic framework for business-government dealings in the United States. It prescribes the formal, public, triallike hearings used by OSHA and other U.S. agencies. But the act provides only rough scaffolding. Subsequent court and agency decisions have created an elaborate, sometimes rococo structure of legal requirements and procedures. Public hearings, like those OSHA held on VC, have evolved into public confrontations with aggressive cross-examination. Lawyers, as specialists in adversarial confrontation, are consequently both the cause and effect of adversarial decision making. The complex legal territory where business and government meet cannot be traversed without them. But their extensive participation has led to ever more complicated rules and regulations that define and delimit business-government dealings.

The effects of a weak executive and a strong judiciary can be especially great in the early years of a business-government relationship, simply because there are so many unresolved issues, large and small, that must be ultimately settled by a judicial process, and for which elaborate procedural requirements are devised. Because ostensibly minor issues may set precedents affecting later issues of greater magnitude, they become battle lines for extensive conflict. This has been especially true for the new regulatory agencies. Their agendas of open issues have been amplified by the relatively long, complicated legislation that created them. As noted before, these detailed statutes were written so that the Congress could oversee highly technical decisions and so that the new agencies would not be captured by their clientele. But the long, detailed legislation created many questions of procedure and substance that had to be settled, usually in court, early in the life of the new agencies. The result was often delay, frustration, and further deterioration of the already fragile, often hostile, relations between companies and new agencies. Businesses gain further evidence of what they consider to be the inefficiencies of government, the ineptitude and antipathy of civil servants, and the ensnarlment of their operations in government red tape.

Because judges make the final decisions, the parties involved have little stake in the proposed regulations put forward by an agency. Instead, the proposal becomes a target for criticism by industry, labor, public interest groups, and others. These groups have not drafted the proposal, they have no direct influence on its contents, and hence can say what they please about it. At public hearings, they escalate their criticisms and hector each other for the benefit of their constituencies. Even the agencies themselves take constituency-pleasing, often adversarial public stands. During the 1970s, OSHA officials could make declarations like that of John Stender, the assistant secretary for OSHA at the time of the VC problem: "We don't equate lives with dollars and cents—not at the Department of Labor." Agency officials know that courts will ultimately tell them whether the balancing should be done and how it should be done. Such public positions, like those of the other parties, discourage compromise and cooperation. And, once the courts have made their decision and created winners and losers, the losers are likely to redouble their efforts to avoid a similar outcome on subsequent issues and will use what is an intrinsically adversarial decision process with greater vigor in the future.

In all the cooperative cases, the important decisions about VC exposure levels generally found expression as informal guidance or technical recommendations not open to challenge in court. Even if the British had issued an approved code of practice for VC, it could have been appealed only under quite restricted conditions. The German VC standard could not be appealed, and the French issued their regulation long after the problem was solved, so the limited possibilities for appeal did not matter. And one scholar has described the freedom of Japanese civil servants from judicial review by writing that, "Rather than a rule of law, a rule of bureaucrats prevails." In all four countries, centuries of strong, legitimate executive branch authority preceded establishment of their generally modest mechanisms for judicial review. Furthermore, with agencies a century old, many of the issues requiring judicial review are settled. The British courts, for example, rendered their major decision on the cost/benefit issue in 1949 and required agencies to consider costs in reaching health and safety decisions.

In all five countries, the basic legislation covering occupational health and safety was far from specific on critical issues. In the OSHA Act, the Congress required that standards be "feasible." In Great Britain, they had to be "reasonably practicable," and in West Germany, the criterion in the Industrial Code was that "employees are afforded all such protection against hazards to safety and health as the nature of the undertaking permits." Laws are vague on these critical issues partly because legislators prefer to delegate these awkward problems to executive agencies. But the main reason that legislators delegate extensively to agencies is that modern industrial problems raise technologically, politically, and managerially complex issues that cannot be resolved once and for all by a few lines of legislation. Therefore, some other institution must exercise final judgment. In cooperative relationships, the institution is an executive agency of government.

The result is a very different set of incentives for the parties affected by government decisions. It is only a partial exaggeration to say that, in the cooperative cases, the government granted the parties the opportunity to influence the VC decision. The parties knew that if they could not agree, the Ministry of Labor would decide the issues and its judgments would almost surely stand. Since the parties had just one opportunity for direct influence, they had strong incentives to use it effectively. In the cooperative cases, important meetings were held in private, often with government representatives in attendance. This reduced the opportunity for constituency rallying and politically motivated hyperbole that could offend the government officials who in the end may make the final decision. When firms have what one Japanese industry official called an "everlasting relationship" with government, they are naturally careful because misbehavior on a current issue could handicap their future dealings with government. Moreover, agencies staffed by career civil servants develop institutional memories, further enhancing their capacity to reward or punish company behavior. On the other hand, firms who face a weak agency led by short-term political appointees, as in the United States, may be more likely to behave in short-term aggressive and opportunistic ways.

The executive agencies in the strong government hierarchies had a repertory of major and minor instruments through which they could exercise their power and the context in which working parties made decisions. Not only did they have the right to set the final regulations, and to do so in an almost unilateral way, they could also determine whether to hold public hearings, what the timetable for working groups would be, how much pressure to exert on these groups to reach agreement, and how and when they would release information to the public and to parliaments on the procedings of the working parties. OSHA, on the other hand, was tightly constrained by law and regulation on all these matters. For example, when OSHA advisory committees meet, the Federal Advisory Committee Act requires that verbatim transcripts be kept and be made available to the public, while in Great Britain, the Official Secrets Act restricts public and media access to a wide range of government documents. The executive agencies were also gatekeepers who determined who could participate in the collaborative deliberations.

In short, then, companies cooperated with government on the VC issue because they confronted stable, respected, experienced, and powerful agencies. Lacking other means of influence, they chose to collaborate. Yet this way of explaining cooperation presents a paradox. The Ministries of Labor in Europe and in Japan had the power to make a unilateral, binding decision on VC, but they did not do so. Rather, government relied patiently on a decentralized, informal process that involved extensive influence and participation by the parties affected. The much stronger executive agencies of government in Europe and Japan consulted and cooperated much more extensively than OSHA, their weaker counterpart.

Industry Hierarchy and Cooperation

The explanation for this paradox partly rests upon the political implications of industry structure. The VC-PVC industries in Europe and Japan were organized into much stronger hierarchies than in the United States. The American industry was essentially a multitude of competing, uncoordinated, or at best, loosely coordinated companies. Each was an autonomous decision-making unit, and each acted more or less on its own, except for the brief and substantially unsuccessfully attempt, under the aegis of the Society of the Plastics Industry, to organize a common industry position. This pattern appears frequently in American business-government relations: the best that industries often achieve is a loose, temporary alliance plagued by internal friction. In the four cases of cooperation, the VC-PVC industries were much more cohesively organized. There was greater congruence among company statements and positions, and there were definite hierarchical or superior-subordinate relations among firms. This was clearest in Great Britain and France, where one or two very large companies dominated the VC-PVC industry. Taken together, the five case studies suggest two broad propositions about institutional arrangements and business-government cooperation. First, intraindustry cooperation, especially in the form of a strong industry hierarchy, makes industry-government cooperation more likely on complex, controversial issues. Second, and more generally, a full understanding of business-government relations on a particular issue must include an account of intraindustry relations. To assume that all companies in an industry are rational, perfectly informed, independent profit seekers who share monolithic views, is to assume away critical institutional arrangements within an industry that can encourage cooperation or conflict with government agencies.

There are practical issues involved. If an industry is concentrated, and if rival firms have collaborated with each other and with government in support of national economic or military objectives, then collaboration on a particular issue—like the VC problem—can occur within the industry much more easily. The three European cases show VC-PVC industries that were much more concentrated than the American industry. Both theory and empirical data strongly suggest that the costs and difficulties of organizing the group grow as the number of individuals in the group increases. In *The Logic of Collective Action*, Mancur Olson proceeds deductively from the assumption of rational, self-interested, perfectly informed actors, to the conclusion that, for large groups of such actors, full and strong support for group objectives is unlikely. This is because many of these rational actors will choose to become "free riders," who take no action, incur no cost, but benefit from whatever action other firms take. Olson concludes that all the members of large groups will pursue joint objectives only through coercion or special inducement, but for much smaller groups—such as relatively concentrated industries—minimal coercion and few special inducements are required.

Empirical confirmation of this conclusion comes from studies of groups ranging from California city councils to the Algerian National Liberation Front.[31] The studies indicate that smaller groups display greater solidarity, consensus, and interaction than larger ones. In the three European countries, a very few firms had done business in the VC-PVC industry for decades. A meeting among a few individuals in a small room was sufficient to represent all the companies directly affected by the VC problem. In contrast, the initial meeting of the U.S. VC-PVC industry at the Society of the Plastics Industry involved more than sixty participants from more than twenty companies. Many participants had never worked together beforehand. Furthermore, the European and Japanese VC and PVC companies had histories in which collaboration commingled with competition. When such firms face government action, their experience with joint effort helps them to anticipate or manage the potential, competitive effects of government action.

The strong executive agencies in the cooperative cases also encourage the formation of strong industry coalitions. The prospect of dealing with a powerful government motivates companies to develop mechanisms for developing joint positions within an industry. For example, the Confederation of British Industry (CBI) was formed in 1965 from three other national industrial organizations to create a single body that could speak on behalf of industry in dealings with government. The head of one of the three associations that merged into the CBI had argued:

> The question is, are we strong enough to speak with authority—to the government, to the unions, to the public, and abroad—in the second half of the twentieth century? The answer, at the present time, is no.

Companies facing a strong government have another incentive to collaborate: to restrain the actions and statements by individual companies that could prompt government action penalizing the entire industry. One official of a large British chemical company said that his industry had occasional difficulty curbing the activities of "cowboys and bully boys" among smaller companies, whose behavior could bring adverse publicity and further regulation. Furthermore, a strong government has interests of its own in dealing with a well-organized industry. The British Labor government encouraged formation of the CBI because it wished to discuss industrial affairs with a single body, and not with three minor ones. Also, the next chapter shows that when government and industry share political responsibility, and when networks blur their boundaries, government depends on industry to provide certain information or perform certain tasks. If industry fails, and if adverse publicity ensues, government agencies, along with the industries with which they collaborate, are held accountable in the press and in parliament. Hence, the agencies seeking cooperation have interests in effective collaboration among firms.

The final factor that encouraged stronger industry coalitions in the cooperative cases is that the basic decisions on VC were made incrementally over a long period of time. Consequently, companies did not have to make specific public commitments about reducing VC levels to unprecedentedly low levels within a very few months of the B.F. Goodrich announcement. Precisely such a requirement led to the unresolved disagreement among U.S. VC-PVC firms on the cost and feasibility of cutting VC exposure levels. By proceeding step-by-step, companies in Japan and Europe had time to gain information about cost and feasibility and had time for discussion and bargaining among themselves. In the United States, the VC and PVC companies had little time to act and they lacked experience collaborating with each other, so each pursued a separate course, and at the OSHA hearing one firm's testimony often undermined the credibility of another's.

The VC cases also indicate that the largest companies—ICI and BP Chemicals in Britain and Rhône-Poulenc in France, for example—are likely to dominate small industry groups, thereby creating hierarchies within the groups and making them even more cohesive. The very large firms could assert with authority that certain approaches to a problem like VC controls were either practical or impractical, and governments and other parties were more likely to accept these views. These very large companies had much more extensive contact with government agencies and much more influence on government agencies and politicians than the smaller companies in an industry. Extensive high-level, ongoing relations with government agencies provide even greater leverage for the very large companies, when their channels of influence have been amplified and refined through elaborate intraindustry networks like the keiretsu in Japan. Moreover, the limited opportunities for judicial review in the three European countries and in Japan further enhance the power of large companies: small companies that object to an industry position or to an agreement between industry and government cannot go to court and challenge the agreement. And lacking the scientific resources of industry giants and the credibility of well-established powerful agencies, they would find it more difficult to challenge these agreements on scientific grounds in other forums, such as the media or in parliament. So a clearly hierarchical arrangement is likely to emerge in which a few large companies have a preponderance of power, can manage and moderate intraindustry politics, and can guide the development of joint industry positions in directions they prefer.

But even if the large VC and PVC companies in Japan and Europe had many practical advantages in reaching industry agreement on shared objectives, this did not by itself guarantee cooperation with government. The companies could have agreed to resist government action. Whether cohesive industry hierarchies encourage business-government collaboration or not depends on the objectives and behavior of the firms that dominate the industry and on the way government responds when confronted with a well-organized industry.

The dominant firms are likely to take moderate positions and seek compromise for several reasons. In the first place, studies have found that because very large firms have much more diverse and complex sets of political and business interests, they tend to moderate their positions on any particular issue.[32] Their executives also have more experience dealing with and compromising with a wide range of interest groups both inside and outside their firms. These considerations were especially important in the three European countries where the VC-PVC producers were subsidiaries of some of the very largest companies in each country. As previously noted, at the time of the VC problem, British Petroleum and Imperial Chemical Industries were the two largest companies in Great Britain (as measured by sales); Rhône-Poulenc Industries was the eleventh largest company in France; and Hoechst and BASF were the third and fourth largest companies in West Germany in 1976. All these companies sold a widely diversified range of chemical products. The huge size of these companies also meant that the effects of the VC decision upon their overall financial performance would be less than for smaller companies or for chemical companies that concentrated on just a few products.

Moreover, just as a large, diversified company may take more moderate positions because of its more complex interests, so an industry, if it has been stably and cohesively organized for a long time, may tend to consider each particular issue, such as VC, against the background of a wide range of other present and future concerns. Without an industry hierarchy, members of a loose, temporary alliance of companies are players in what are essentially zero-sum games in which one company's gain tends to be another's loss and industry's gains tend to be the government's loss. A stricter VC standard, for example, benefits companies with greater volume of output, over which they could spread the fixed costs of complying with the regulation, and companies with greater technical sophistication. A stricter standard also tends to benefit workers at the expense of company profits. And in political terms, a government's victory, like OSHA's victory in the VC case, tends to be an industry's loss. If, however, companies have long-term relations with each other and with government, extending across a range of issues and over a number of years or decades, there are far more opportunities to make complex trade-offs and adjustments within the industry hierarchy and between industry and government. In effect, there are institutional arrangements that facilitate the conversion of conflict-ridden, short-term, zero-sum games into long-term negotiations and collaboration.

Strong trade associations also encourage moderate industry positions. Senior association officials often share educational and social backgrounds with their government counterparts, and they work with each other on a wide range of issues, year after year. The associations are also enmeshed in a network of issues spanning a variety of subjects and a number of years; consequently, association officials behave in ways that reflect the opportunities for complicated trade-offs and adjustments across the full range of issues linking the associations and government. In this study, the clearest example of such an institutionalized, short-term link was the practice of the Japanese PVC Association of having as its vice-chairperson a former official of the Ministry of International Trade and Industry.

When companies and their associations are well organized, they also have much less need for adversarial rhetoric and hyperbole (such as references to industry shutdowns or government regulatory czars) to move companies to act. In fact, one of the functions of the bombastic public rhetoric in adversarial relationships may be to encourage the mobilization of parties that might otherwise remain neutral. It was only after OSHA announced its proposed zero ppm standard for VC—the "bombshell"—that U.S. industry began its serious, though ultimately unsuccessful, efforts to organize. Adversarial rhetoric is in some degree a surrogate for the factors that encouraged intraindustry collaboration in the cooperative cases.

In the United States there was no strong industry hierarchy to encourage collaboration on the VC issue. Nor was there a strong trade association with extensive experience dealing with government. In fact, the initial stages of the plastics industry response to the VC problem involved intense jurisdictional conflict between the Manufacturing Chemists Association and the Society of the Plastics Industry. Neither was certain it was clearly responsible for or that all its members would benefit from active involvement in the VC issue. Furthermore, at the time the Society of the Plastics Industry took on the VC problem, it had very little experience dealing with OSHA on occupational health and safety matters. Thus, a weak association, uncertain about its role, lacking relevant experience with OSHA, had to attempt to coordinate the activities of a highly fragmented industry in a short period of time and do so on what became a zero-sum issue for the parties concerned.

In general, when industry and government hierarchies are strong, cooperation is more likely. Yet there remains a curious partial exception to this proposition—the approach taken by the Japanese to the VC problem: a large, competitive plastics industry managed to cooperate quite successfully on the VC problem. Of course, the Japanese VC and PVC companies may have been predisposed to cooperate by virtue of a national industrial history involving extensive collaboration among companies, especially those within the large economic conglomerates. Furthermore, the Japanese may indeed have the cultural predisposition toward collaboration and consensual decision making so widely attributed to them. Also, two-thirds of the Japanese VC and PVC companies had been involved at least indirectly in the Minimata problem, which gave them previous experience in dealing with a critical environmental issue. Moreover, as noted, Minamata disease shared similarities with angiosarcoma of the liver induced by VC exposure. Therefore, these companies, regardless of cultural predispositions, had incentives of the highest order to collaborate with each other and with government to avoid a recurrence of the deep suspicions, harsh publicity, and strong, almost punitive government actions that the Japanese government took in response to the Minamata mercury poisoning.

Even so, the Japan PVC Association and the industry it represented do not fit precisely into the pattern of strong hierarchies that characterized the European VC-PVC industries and distinguished them from the American. A full account of how institutional arrangements encourage cooperation, and a full explanation of the collaboration in Japan, must involve an additional factor: the elaborate, informal, institutionalized networks of personal contact linking industry and government. The networks that span the industry and government hierarchies created powerful incentives to cooperate in Japan as well as in the three European countries. They also complete the explanation of the paradox described earlier: that powerful government agencies with the authority to act virtually unilaterally chose to cooperate with companies rather than to impose a solution upon them.

Networks and Cooperation

In the VC case, the networks linking major institutions strongly promoted cooperation because they permitted, encouraged, and legitimated the *direct* participation of all major parties in the resolution of the problem. This was true with the bilateral relations in Japan, the trilateral relations in Great Britain and France, and the formal multipartite relations in West Germany. The opportunity for direct influence on high-stake decisions was a crucial factor in explaining cooperation. This was strongly suggested by the U.S. experience with the VC problem, in which the major parties could exercise only indirect influence on the government decision. A preliminary decision was made in secrecy by agency officials, and then a final decision was made—again in secrecy—by judges. Lacking the opportunity for direct influence, industry and labor saw indirect means of influence, and these were intrinsically adversarial.

The adversarial character of public hearings, such as the OSHA VC hearings, and of the judicial proceedings that regularly followed decisions of American government agencies has already been discussed. But industry and labor seek to influence government decisions in other ways as well. Since the stakes in many of these decisions are very high—hundreds of millions of dollars, the lives and welfare of workers and communities, the personal reputation of major participants, and the informal and formal precedents that will later bind the activities of a new agency—industry and labor bring to bear all available means of influence on a decision; when they do so adversarial behavior ensues.

The principal means of indirect influence are White House intervention, congressional intervention, and media attention. All these institutions—the White House, Congress, and the media—have limited resources and a wide range of issues on which they may focus their efforts. To secure their attention and to stimulate action, a group needs a simple, clear, and commanding message; it needs to show that many constituents, readers, or viewers share its agenda. And when political pressure is necessary, a message must stimulate powerful groups to mobilize. Thus, before major environmental decisions in the United States, industry raises the banner of multibillion-dollar losses, while trade unions and public interest groups, and on occasion, agency officials announce forthcoming epidemics of dreaded diseases. Ultimatums, hyperbole, and the spotlighting of dramatic episodes, like Love Canal, are especially helpful for mobilizing constituencies, as are adversaries, such as predatory corporations, monopolistic trade unions, regulatory czars. To stir opposition to an OSHA reform bill, for example, trade unions distributed posters urging, "kill the Schweiker bill before it kills you." Agencies can also

use similar appeals to public opinion to secure their independence from other parts of the government, such as the White
House or Congress. In the early years of the Environmental
Protection Agency, William Ruckelshaus took vigorous measures against a number of very prominent American companies, partly to display the independence of his agency and
partly to mobilize constituencies outside government in its
support.

To exercise some form of indirect influence, a group must
make itself heard above the ongoing cacophony generated by
hundreds of other groups seeking congressional, White
House, or media attention. Moderate, reasoned, tolerant
statements are far less potent than adversarial sound and fury.
Frustration that arises in this situation can lead to bizarre
behavior. The mixture of high stakes and constrained opportunities explains, for example, the fact that adult Americans who
are highly paid partners in prestigious law firms and members
of America's socioeconomic elite, can often be found feeding
coins into pay telephones in the drafty lobbies of government
office buildings just before major regulatory decisions are announced. In the VC case, an employee of a public relations
firm hired by the VC and PVC companies stood in the back of
the room in which OSHA announced its final decision for
VC. As soon as the level was announced, he signaled an
attorney at the end of a long corridor, who, in turn, signaled
another attorney who had been feeding quarters into a pay
telephone to keep the line open. At the other end of the line,
yet another attorney was waiting at a federal courthouse in
New York City to file appropriate papers seeking appeal of the
VC decision. This procedure enabled the Society of the Plastics Industry to file for a review of the VC standard within two
minutes of its announcement. Known as "the race to the
courthouse," this practice is yet another way in which parties
adversely affected by a government decision, and frustrated by
limited opportunities to influence it, seek to exert some indi-

rect influence—in this case by strategically selecting the court that will hear their appeal.

In the three European cases, the intricate networks, institutionalized as quangos, enabled industry, labor, and government directly to influence the VC standard. They were also jointly responsible for setting and implementing the policies of a major national health and safety agency. Within this institutional framework, multipartite working groups—which were themselves ad hoc, miniquangos—made regulatory decisions. The first consequence of this direct participation was that all parties had much stronger stakes in the final VC standards. They shared the legal responsibility and, inevitably, the political risks of their decisions. In the event of a highly publicized health or safety problem, it is more difficult for a trade union or industry association to accuse health and safety agencies of dereliction of duty when unions and management are directly responsible for overseeing the activities of inspectors. As a consequence, even in Japan, where there were no quangos, the result of joint decision making and direct influence was that all parties shared political responsibility. In the VC case, the shared political accountability among the Japanese may have been even greater than in Europe because of the strong parallel between Minamata disease and VC-related liver cancer. If Japanese industry and government had failed to achieve an effective resolution of the VC problem, both could have found themselves cast in a very unfavorable public light. Japanese labor and management also shared a responsibility for the VC standard. They conducted a joint mission to Europe and a series of plant visits in December 1974 and February 1975 to develop a common data base, and they participated, along with members of the Ministry of Labor, in the two major tripartite meetings at the Ministry of Labor, which approved the final standard. The government was also linked to the companies through the two subsidies it provided. In all the cooperative cases, there was extensive direct participation in critical decisions that made the major parties political bedfellows who shared political responsibility.

Direct participation also encouraged cooperation by setting in motion several self-sustaining cycles. For example, in a cooperative relationship on any particular problem, there is an opportunity for large companies, their trade associations, and labor unions, to expand and strengthen their informal contacts with government agencies, and thereby enhance their political power. In the United States, industry viewed the VC problem as a risk rather than an opportunity. This is why the Manufacturing Chemists Association decided not to become involved in the VC issue, and it is also the reason the Society of the Plastics Industry was at first reluctant to act. There were no long-term ties between business and government, so the VC episode provided no chance to build political capital.

Shared responsibilities also strengthen the credibility and legitimacy of decisions that European and Japanese multipartite working groups make. Such groups rely heavily on information provided by workplace inspectors, who generally have years or even decades of experience in the problems and operations of the industries for which they are responsible and whose managers are respected career civil servants. All these factors raise the quality of the information that the inspectorates develop on cost, compliance, and technological alternatives. And, just as important, the parties involved with a decision, being basically external groups, are more likely to perceive this information as objective and accurate because it is not prepared on behalf of any particular group.

In the VC episode, the multipartite working groups also relied extensively on experts from the industries that would be directly affected. These experts, who themselves often had decades of experience with VC and PVC operations, were direct participants at all stages of the development of VC regulations. Furthermore, especially in Germany, outside scientists were direct participants in working group discussions. For example, the BG Chemie official responsible for handling the VC problem belonged to its technical inspectorate and had already conducted a two-year study of the health risks of VC-PVC production in West Germany. He was unambiguously an expert in the VC-PVC industries, but he conducted studies and investigations on behalf of the BG and its multipartite VC working group, both of which were closely supervised by joint labor-management groups, and by the West Germany Ministry of Labor. In contrast, in the United States, the principal role of experts was to testify briefly at OSHA public hearings. Each party then generated its own "objective" data, made its own independent analysis, and then sought to achieve credibility through efforts that discredited the data and judgment of the other parties. All information, accurate and inaccurate, was tarred by the same brush.

Multipartite groups and bipartite networks also legitimate the informal, private, cooperative approach to decision making through the labor unions' participation. On health and safety problems, the interest of labor generally lies in greater protection. Thus, for VC, labor representatives in all the working parties pressed for lower and lower exposure levels. Their effort countered industry concerns about costs, profit, and competitiveness. As a result, both sides of the cost/benefit issue gained full and vigorous representation in the working parties' deliberations. Furthermore, quangos, government inspectorates, and health and safety committees gave labor unions access to technical expertise that they otherwise might not have had. This reduced the likelihood that workers would be weak, third-class participants in the multipartite deliberations. The availability of this expertise also aided the government in the working groups. Naturally, agency officials cannot have expertise in the technical intricacies of a wide range of industrial operations, and so they also benefited from the specialized advice provided by experienced inspectors.

The networks linking the major parties and, in particular,

the working groups facilitated the incremental approach to reducing VC exposures that appears in all the cooperative cases. The direct consequence of this incremental approach was less uncertainty and less antagonism by the industry officials whose operations would be critically affected by the VC regulations. Companies did not have to make early public announcements, as they did in the United States, about the cost and feasibility of drastic, unprecedented, and controversial reductions in VC levels. Instead, in all four cases of cooperation, the companies could make broad commitments to vigorous effort and agree on interim targets that seemed feasible. All parties knew that labor unions and government officials, acting through the multipartite networks, would press companies to cut VC exposure. It was also clear that companies would have very limited opportunities to appeal whatever the final decisions were. And furthermore, the experienced inspectors, with wide access to company documents and operations, could confirm independently industry's efforts to reduce exposure and the costs that industry was incurring. Thus, it proved practical and legitimate to reduce VC exposures in a gradual, stepwise fashion and periodically assess each step and the feasibility and cost of subsequent steps. Clearly, the informal, flexible, and incremental approach to assessing cost and benefit and to determining the timing and extent of major actions reduced some of the fears and frustrations that provoked American company officials to vigorous adversarial actions. In all the cooperative cases, there was never any question that issues of cost and feasibility would be carefully and continuously scrutinized as the VC problem was resolved.

The careful incremental approach reduced the surprises and frustrations that exacerbate business-government conflict in the United States. As noted earlier, some U.S. company officials responded to OSHA's proposed standard by asking, "Did you see what the bastards did to us today?" In the cooperative cases, it was possible for the parties to avoid the surprise and shock of abrupt official decisions by relying on private discussions and delicate application of what the French call the rule of anticipated reactions. Furthermore, through direct detailed communication on complicated issues, the parties avoid the aggravations that inevitably arise when experts must explain complicated subjects on paper to amateurs (such as judges and the inexperienced, high-level political government appointees at OSHA). This frustration grows when such communication must take place under circumstances of crisis and under a constraining web of procedural requirements.

The networks also provided an institutional umbrella under which personal relations could develop among members of the working parties as a consequence of their small group activities. The representatives of various parties had many opportunities to patiently explain their perspectives on the VC problem and to come to understand each other's viewpoint. This was especially important since members of the working parties represented groups with different attitudes and priorities on the medical, economic, and political issues at stake. Because the approach to the VC problem extended over several years in the cooperative cases, there was ample time for personal relations and shared perspectives to develop.

The small working group also enabled the parties to speak directly about complex issues, rather than rely on the giant documents that are routine features of U.S. regulatory decision making. Representatives of industry had ample opportunities to make suggestions and press for alternatives they considered reasonable. The entire pattern was a more gentle, less intrusive way for government officials to involve themselves in the efforts of industry to reduce VC exposures. This surely encouraged cooperation, as did the fact that the inspectorates were acting primarily as advisers. Consequently, plant managers were more likely to cooperate with them than they would in the United States, where there were legitimate concerns that candor and cooperation could be self-incriminating and could lead to citations, fines, or possibly even imprisonment.

Networks linking the major parties also encouraged cooperation in a final way. Because much of the decision making and discussion was informal, there were far fewer opportunities for parties to engage in legalistic delays, alleging that other parties have violated some procedural requirements. Since the parties relied on informal communication and quasi-formal authoritative guidance, they did not face the problems and frustrations of trying to use simple, legal instruments and rules to specify solutions to extremely complicated problems that can vary from company to company and even from plant to plant. Nor did they encounter the frustrations and problems that arise when highly specific and legalistic requirements produce unintended consequences, which further frustrate the parties, create greater hostilities and apprehension, and ultimately require additional procedural requirements and reviews.

Legitimate, informal, flexible networks complete this account of how institutional arrangements can encourage business-government cooperation. Strong government hierarchies, taken alone, are an inadequate explanation because they would suggest that cooperation springs from superior power. But this would be coercion, not cooperation: victims who obey kidnappers are not partners in a genuinely collaborative relationship. Moreover, an account of business-government cooperation based solely on superior government strength would not explain why powerful agencies choose to collaborate rather than issue unilateral decisions. Nor is the combination of strong government hierarchy and strong industry hierarchy a satisfactory explanation, since the two powerful sides could become adversaries who fought or coexisted under conditions of cold-war hostility. Networks, along with strong hierarchies, must play a role in explaining cooperation on complex controversial issues because they render business and government mutually dependent and create a set of incentives and opportunities favorable to collaboration.

Chapter 8
IMPLICATIONS

Overall, the VC case shows that adversarial and cooperative basic relations are two quite different ways of accomplishing critical political tasks. Each displays a different pattern of legitimacy, accountability, and participation. But the major difference is that cooperation accomplishes this task in ways that help the performance of analytical and managerial tasks. Cooperative decision making can move incrementally on the complex and uncertain problems created by modern industrial production; its varied, informal, semiprivate, and flexible networks are superior channels for communicating information about technically complex, highly controversial subjects. In addition, cooperation involves a decentralized and legitimate reliance on industry experts rather than on scientific generalists and legal personnel.

Does the VC episode imply that Americans should seek
greater cooperation between business and government on
other complex, controversial issues? And if so, what would be
useful steps in this direction? The five cases suggest several
ways of thinking about these important questions.

Perhaps the most important implication is that cooperation
is a special-purpose tool, one that will work in varying degrees
for some problems but will prove useless or even harmful for
others. In the VC episode, cooperation displayed clear managerial and analytical advantages. Strong hierarchies, linked
by elaborate networks, helped companies and agencies to
gather, assess, communicate, and legitimate data and judgments on a problem of high technical complexity. But the VC
problem was unusual—precisely because public and private
officials *could* make so much progress simply by developing
medical, scientific, engineering, and cost data. By proceeding
incrementally, and by following rather than forcing industry
efforts to control VC, the working parties pared down some
critical uncertainties. In time, they learned that the VC
hazard was very serious, that engineering changes could cut
exposures to single-digit ppm levels, and that companies could
make the changes without financial havoc.

For several reasons, however, such joint problem solving
will not work nearly so well for many other modern industrial
problems. In the first place, critical information often does not
exist, and cooperation—whatever the institutional arrangements—cannot create it. The industrial policy problem of
promoting "sunrise industries" is a clear example. Business
and government could perhaps work hand-in-hand to nurture
new industries, if they knew what they were. But no one
knows and no one can know—except in the broad sense in
which "everyone knows" that financial services or high technology are the industries of the future. (Conventional wisdom
viewed energy in the same way during the 1970s.) If the critical data could be conjured into existence, they would show

which products, services, and companies customers will prefer in five or ten years. But this is information about the future. In the VC case, the laboratory experiments, epidemiological surveys, engineering experiments, and cost studies—which were wellsprings of critical data—gave perspective only on the past and present. Yet for many urgent economic problems, only the passing of time and the evolution of marketplaces will reveal critical information. In theory, of course, business and government could seek some cooperative way of assessing which goods, services, and products were most likely to succeed. But in the United States, companies and capital markets already try intensely to do just this. Business-government cooperation to generate further data would run a strong risk of redundancy. The VC problem was quite different. Much of the data needed was precisely the sort that cooperation could develop, assess, and legitimate. Under these conditions cooperation proved useful.

The VC case also suggests that cooperation to aid so-called "sunset" industries would encounter another information problem. The parties affected by a declining industry usually have abundant data on the causes and consequences of their common plight and the alternatives facing them. Representatives of companies, government agencies, labor unions, and communities know each others' interests, positions, and values. They share powerful incentives to find some way of resolving their common problems: the stakes include thousands of jobs, millions of dollars of tax revenue and company profit, tens of thousands of votes. In these circumstances, it is difficult to see what good would be accomplished by cooperative efforts to gather even more data and assess it. The problem is not simply data, and multipartite cooperation along the lines of the VC case would not be a useful tool.

The VC cases also suggest that cooperation is a special-purpose tool in a second sense: it is probably much better suited to questions of means than to questions of ends. Members of the VC working parties in Europe and Japan all agreed from the very beginning on the goal of cutting VC levels dramatically. Given this objective, they concentrated on what were essentially technical questions of implementation. Of course, the distinction between ends and means was not airtight. As companies reached each interim objective for VC exposure, a basic question recurred: should the companies make further efforts, or had the basic objective been accomplished? In this attenuated form, the question of ends did persist. But all parties had agreed to reduce VC level to the lowest feasible level, and it was established practice to consider the health of the industry in deciding what was feasible. Cooperation succeeded because, for the most part, the working groups concentrated on the question of how to meet their objectives, not the question of what their objectives should be. Cooperation thrived on a narrow issue of implementation.

But many other modern industrial problems are Gordian knots, intertwining difficult questions of means and ends. They pose fundamental questions about the rights of citizens, the role of government and private enterprise, and basic human values. Such issues have no circumscribed question of technical fact that cuts through all their complexities and uncertainties, political and technical. This suggests that cooperation is better suited to simpler problems in which the task is to translate clearly defined and widely shared objectives into action.

The third limit on business-government cooperation is the difficulty of arranging effective participation by the parties directly affected by a problem. Without their involvement, a cooperative effort—however well intentioned—and its outcome, will not be considered legitimate, and excluded parties may try to frustrate the implementation of whatever decisions are reached. Moreover, there is also the possibility that the parties who do participate will collude against the interests of excluded groups.

The cooperative VC deliberations avoided this pitfall because all the parties directly affected by the problem participated directly in all the important discussions and decisions. Moreover, two of the parties, industry and labor, pursued conflicting objectives on critical issues. The union officials, on the whole, gave higher priority to protecting workers' health, while company officials had strong interests in the economic health of their firms and their industry. As a result, a "micropluralism" of countervailing pressures, moderated by powerful government officials, kept the two parties from self-interested deal making. Furthermore, the institutional arrangements favoring cooperation did not permit any parties to suffer a kind of second-class citizenship and participate only in a formal or ritualistic way. Workers' representatives could have suffered this fate for, while they did have considerable political leverage, they lacked expertise on the complex technical problems raised by VC. Government officials ran the risk of a similar handicap. But the networks of health and safety committees and the efforts of powerful, experienced inspectorates corrected an imbalance of technical prowess that favored the huge chemical companies. Government and labor union officials do not accept the judgments of company experts simply because the industry representatives are well intentioned, likable people, with whom they have become friendly. It is the institutionalized opportunities for independent corroboration, by parties with interests differing from those of industry, that make reliance on industry judgments legitimate and credible.

But for many issues, especially economic issues, such micropluralism may prove elusive. The problem is the difficulty of creating arrangements that enable consumers to represent their interests. Markets do "organize" consumers, but not in ways likely to enable their successful representation or participation in multipartite meetings. Unlike labor, management, and government, consumers are disorganized. Yet without consumer representation, business-government cooperation could do more harm than good. The familiar hazard is that organized bodies like companies, labor unions, and government officials will negotiate anticonsumer schemes of subsidy and protection. If anything, the very existence of a strong industry hierarchy—one of the institutional arrangements that encouraged cooperation in the VC case—is a possible symptom of an industry dominated by a few giants who may have lost the habit of competition.

This limit on cooperative problem solving by business and government extends beyond economic issues. Inability to arrange effective participation could also enfeeble joint efforts on occupational, environmental, and social issues. Decades after the rise of powerful labor unions, most workers in the industrial West are not union members; even in Great Britain, the most highly unionized of the five countries discussed, only one worker in two belongs to a union. Furthermore, many unions are weak, and many others have other priorities, such as pay and job security, that exhaust their limited resources. Many industries consist of myriad small companies. Their trade associations often exist in name only, and sometimes several associations compete intensely to represent the same companies.

Given these limits, how attractive is business-government cooperation for resolving complex, controversial problems? The answer is that—even as a special-purpose tool—cooperation is quite attractive. In part, this is because it can be, as the VC cases show, a very powerful instrument for simultaneously achieving analytical, managerial, as well as political goals, if the key parties are effectively represented, if the main task is implementing clearly defined and widely shared goals, and if gathering and assessing information can, indeed, reduce critical uncertainties. Since almost all government agencies face some problems with these features, this powerful tool should be in the repertory of problem-solving approaches available to agency officials. Even as a special-purpose tool, cooperation should find many important uses. In a regulatory state, over-laid upon a welfare state and a mixed economy, the activities and interests of business and government commingle across a vast territory of issues, problems, and opportunities. Conse-quently, the natural agenda for business-government coopera-tion, though limited, remains quite substantial.

The second broad lesson that the VC cases suggest is that Americans should view the task of promoting greater coopera-tion as a long-term effort, one likely to require decades rather than years. Barring a catastrophic crisis, such as the Great Depression or the two world wars, more cooperative relations can be built only through patient, step-by-step efforts across a range of institutional fronts. Viewed from afar, cooperation—at least in the VC case studies—was encouraged by strong hierarchies linked by elaborate networks. But these overarch-ing patterns are actually accretions of particular institutional arrangements that jointly shape the behavior of public and private officials. For example, business and government in Europe and Japan were free to hold discussions in private, unimpeded by "sunshine" kinds of laws enacted in the United States. National civil services offered positions of sufficient authority and status to attract and keep many of the most capable graduates of the best universities. Senior ministerial positions were open to the most talented and experienced civil servants, and not to political appointees who quickly enter and exit through revolving doors. Government officials could draw upon a graduated range of instruments of administrative guid-ance—formal regulation, technical guidelines, codes of prac-tice, and so forth—and they had substantial discretion in choosing when and how to use them. On the industry side, company officials do not break the law if they discuss and negotiate in private with government officials. And they run much less risk of law breaking when they talk with other com-pany officials. Antitrust laws, and their customary enforce-ment, permit much fewer extensive joint efforts among mem-

bers of an industry. Opportunities for judicial review are much
more limited, and judges are much less willing to overturn
decisions of executive bodies.

Movement in any of these directions in the United States is
likely to be quite slow. Established institutional arrangements
usually serve the interest of powerful groups, who will
mobilize to protect those interests. "Established practice" is
also a kind of glacier—aggregated of many laws, precedents,
standard operating procedures, traditions, habits, and conventional wisdom—moving with steady, powerful inertial force.
Moreover, established practice is often sound practice, reflect-
ing a society's fundamental political values and lessons drawn
from past experience. In the end, greater cooperation will
require changes across several fronts, not piecemeal efforts.
OSHA's experience with advisory committees confirms this.
Though the OSHA Act does not require it, the agency has
frequently appointed advisory committees—with members
drawn equally from business, labor, and the public—with the
mandate of recommending how OSHA should regulate a particular problem, such as noise or coke oven emissions. (There
was no advisory committee for VC because OSHA followed
emergency procedures.) One study of the embryonic attempts
at cooperation concluded that "few compromises are struck,
and consensus is hardly ever reached."[33] Committee meetings
are open to the public; attorneys for various parties often at-
tend; verbatim transcripts are kept and published; the committees disband as soon as OSHA takes the next regulatory step of
holding public hearings; and representatives often lack the
authority to commit the groups they represent. In theory, the
advisory committees are an excellent idea. In practice, customary institutional arrangements undermine them.

Do OSHA's fruitless efforts, along with the VC cases, imply
that efforts to promote greater business-government cooperation in the United States are doomed? Perhaps the most fundamental obstacle to greater cooperation is that it may require

much more powerful executive agencies of national government than Americans will accept. A common thread runs through "government in the sunshine" rules, detailed laws limiting agency discretion, wide opportunities to challenge agency actions in court, "hard look" judicial review, and other American practices. It is the aim of limited government power. Yet a strong government hierarchy was the single feature shared by all four cases of cooperation on VC. Labor unions varied from strong in West Germany to weak in France; the European industry hierarchies were stronger than their fragmented, intensely competitive Japanese counterpart. Political culture and national history varied from country to country. But across the board, the Ministries of Labor were very powerful: they had the authority to impose final, binding standards for VC. If cooperation presupposes strong government, then—this argument runs—the American body politic will likely reject transplanted institutional arrangements favorable to cooperation.

But this conclusion is wrong, for the VC cases suggest a more complicated line of reasoning. In principle, the Ministries of Labor did have the power to act almost unilaterally. Yet, in practice, they did not do so. They did not use, much less abuse, their potentially overwhelming power, because their links to companies and unions enmeshed them in a complex web of mutual interdependence. Within this network, government officials chose to move patiently, privately, gently, and unobtrusively. They thereby secured several advantages. Government officials gained full access to industry expertise in arcane areas of science and engineering; they shared the high political risks that pervade modern industrial problems; they resolved a difficult, dangerous problem with minimal frustration, resentment, and hostility for all concerned; their VC standard was more likely to be effective because it was based on expert industry judgment and actual, shop-floor efforts at VC control; and they set a precedent likely to secure similar advantages on future problems. Government officials ran few risks in taking this approach: if cooperation failed, they could have regulated VC on their own terms.

The networks intertwining the strong government hierarchy created incentives and opportunities that discouraged abuses of power that strong government alone might permit. Moreover, government power did not suppress a pluralistic expression of the viewpoints, interests, and values of the parties affected by the VC problem. Instead, a micropluralism of multipartite working groups prevailed; all sides of the VC problem were represented. Government officials in the cooperative cases may have been predisposed toward representing the larger companies and trade unions, and parties ran a risk that they would not be invited to subsequent negotiations if agency officials thought they had not bargained in good faith in the past or had tried to use the working parties mainly to secure media attention or pursue other purely political goals. Nevertheless, all the major views and interests on the VC problem found representation. There was, no doubt, a different pattern of participation and accountability from that in the United States. The cooperative cases practiced pluralism defined in terms of the representation of interest, not in terms of access and influence for a large number of individuals and groups.

In the end, the advocates of greater business-government cooperation in the United States must be clear about the magnitude of what they are recommending. Implicitly, but inescapably, they are calling for different standards of accountability and participation from those now written into law. They are calling for different institutional arrangements between business and government. And they are implicitly advocating different relations within business and government, since one of the lessons of the VC episode is that intraindustry and intragovernment relations strongly influence the prospects for industry-government collaboration. They are not advocating reforms that will simplify business-government relations—if

anything, cooperation is more complex because a wide range
of informal and middle-level activities must be coordinated.
Indeed, the cases of successful cooperation on VC suggest that
the administrative substructure of effective business-government relations will mirror the full range of complexities—
analytical, managerial, and political—of modern industrial
problems.

Above all, promoting cooperation in the United States is
difficult because an abiding Lockean strain in American political life and thought holds that the major institutions of society
should be kept at arm's-length from each other, that their
boundaries should be clearly marked, not blurred by informal
networks, and that the exercise of government power should
be public and subject to judicial review. The canons of legitimacy, and the institutional arrangements that are their progeny, now impede extensive business-government cooperation.
Legitimacy runs counter to effectiveness. Therefore, the first
steps toward greater cooperation involve changes in the ways
Americans think about participation, accountability, and
legitimacy, along with changes in our institutional arrangements.

Building greater business-government cooperation in the
United States, therefore, is a daunting task. But for certain
problems, the effort will yield large rewards. This study of a
complicated, controversial problem vividly displays the administrative, analytical, and political advantages of cooperation: its incremental approach to complicated and uncertain
issues; its heavy reliance on industry expertise; and its ability to
legitimate decisions and joint decision making. Cooperation is
a superior approach—if a problem can be resolved by gathering and assessing data, if it is a problem of means rather than
ends, and if all important interests and views can be effectively
presented.

NOTES

1. "Vinyl Chloride and Cancer—A Study in Prevention," a pamphlet published by the B.F. Goodrich Company, July 1976, 3.
2. "Vinyl Chloride Chronology," photocopy, undated, prepared by the Manufacturing Chemists Association (now the Chemical Manufacturers Association), Washington, D.C., 7.
3. The study is P.F. Infante, "Oncogenic and Mutagenic Risks in Communities with PVC Production Facilities," in *Occupational Carcinogenesis Annals of the New York Academy of Sciences*, ed. Umberto Saffiotti and Joel Wagoner, vol. 271 (1976): 49ff.
4. Alexis de Tocqueville, *Democracy in America*, vol. 2 (New York: Vantage Books, 1945), 263.
5. Nicholas Ashford, *Crisis in the Workplace* (Cambridge, Mass.: MIT Press, 1976), 142.
6. David Landes, *The Unbound Prometheus* (London: Cambridge University Press, 1969), 192.

7. Samuel Eliot Morison, *The Oxford History of the American People* (New York: Oxford University Press, 1965), 731.

8. Ibid., 732.

9. The 500 ppm standard was one of approximately 400 "consensus" standards that OSHA adopted in 1971. These had been developed by industry groups and government agencies before OSHA's creation.

10. This summary of the two studies is taken from Firestone Plastics Co., "The Statement of the Firestone Plastic Co.," 22–23.

11. Bommarito is quoted in Circular Letter no. 30/74, International Federation of Chemical and General Workers Unions, Geneva, 2.

12. "Focusing on Fabricators," *Chemical Week*, 26 June 1974, 14.

13. See *New York Times*, April 1974, 58, and 11 May 1974, 17, for both studies cited.

14. "Brief for the Petitioners-Intervenor," *Firestone Plastics Co. and Union Carbide v. U.S. Department of Labor*, 509 F. 2nd (2d Cir. 1975), 33.

15. Ibid., 39.

16. OSHA, "Exposure to Vinyl Chloride," 39 *Federal Register*, 39892 (1974).

17. Health Research Group, "In the Matter of the Economic Impact Statement on the Proposed Permanent Standard on Occupational Exposure to Vinyl Chloride," Comments Docket no. OSH-36, 14.

18. *Chemical Week*, 6 June 1975, 1.

19. This account of the VC compliance costs is taken from H.R. Northrup, *The Impact of OSHA* (Philadelphia: Wharton School, University of Pennsylvania, 1978), 371–89.

20. H.R. Northrup, *The Impact of OSHA*, 371–89, gives an estimate of the $130 million. An estimate of capital costs of $178 million was provided by John R. Lawrence, technical director, Society of the Plastics Industry, personal correspondence, 24 October 1980.

21. The estimate for Great Britain is from Dr. John Stafford, "Safety Balance: VCM Case Study," *Science and Public Policy*, April 1977, 28 August 1976, 1718; and the West German is from Verband Kunststofferzeugende Industrie e. V. (Association of Plastics Producers), *VC/PVC: Example of a Problem Solved* (Frankfurt am Main: Verband Kunststofferzeugende Industrie e. V., 1975), 18.

22. "Vinyl Chloride: Code of Practice for Health Precautions" (London: Health and Safety Executive, 1975), 2.

23. Ibid., Appendix 1.

24. Association of Plastics Producers,"VC/PVC: An Example of a Problem Solved," a pamphlet, 1977, 17.

25. The account of the response to the findings described at the Tokyo conference is based on "The Vinyl Chloride Problem in Japan," Japanese PVC Association, photocopy, October 1974, 5–6.

26. Memo on the Industrial Hygiene Committee of the Japanese PVC Association, photocopy, September 1981, 9.

27. ORA Number 348, Ministry of Labor, Tokyo, Japan, 1975, passim.

28. The guideline used a geometric mean and standard deviation on the grounds that the distribution of concentration levels of hazardous substances in the workplace atmosphere is a logarithmic distribution and not a normal distribution.

29. The guidelines recommended a standard deviation and not a ceiling level as a way of limiting temporary exposure to high VC levels. Other countries, for example, the United States, used a ceiling level. The reason given in the technical guidelines is that the ceiling approach required extensive, or perhaps continuous, measurement of the workplace, while the standard deviation approach permitted temporary high levels to be monitored without such extensive measurement.

30. Ibid., p. 3 of translation provided by Japanese PVC Association.

31. Robert D. Putnam, *The Comparative Study of Political Elites* (Englewood Cliffs, N.J.: Prentice-Hall, Inc., 1976), 114; and Steven Kelman, *Regulating America, Regulating Sweden* (Cambridge, Mass.: MIT Press, 1981), 146ff.

32. Raymond A. Bauer, Ithiel De Sola Pool, Lewis Anthony, *American Business and Public Policy* (Chicago, Ill.: Aldine-Atherton, Inc., 1972), 152–53.

33. Albert L. Nichols and Richard Zeckhauser, "Government Comes to the Workplace: An Assessment of OSHA," *Public Interest*, Fall 1977, 6.

INDEX

This book was set electronically from the author's word processing diskettes. Editing revisions and typesetting encoding were done in the publisher's office. Electra was the second typeface designed by the renowned typographic artist William A. Dwiggins and was introduced in 1937. This book was set in Electra roman and cursive type and was printed by offset lithography on acid free paper.